Information and Instructions

This shop manual contains several sections each covering a specific group of wheel type tractors. The Tab Index on the preceding page can be used to locate the section pertaining to each group of tractors. Each section contains the necessary specifications and the brief but terse procedural data needed by a mechanic when repairing a tractor on which he has had no previous actual experience.

Within each section, the material is arranged in a systematic order beginning with an index which is followed immediately by a Table of Condensed Service Specifications. These specifications include dimensions, fits, clearances and timing instructions. Next in order of arrangement is the procedures paragraphs.

In the procedures paragraphs, the order of presentation starts with the front axle system and steering and proceeding toward the rear axle. The last paragraphs are devoted to the power take-off and power lift systems. Interspersed where needed are additional tabular specifications pertaining to wear limits, torquing, etc.

HOW TO USE THE INDEX

Suppose you want to know the procedure for R&R (remove and reinstall) of the engine camshaft. Your first step is to look in the index under the main heading of ENGINE until you find the entry "Camshaft." Now read to the right where under the column covering the tractor you are repairing, you will find a number which indicates the beginning paragraph pertaining to the camshaft. To locate this wanted paragraph in the manual, turn the pages until the running index appearing on the top outside corner of each page contains the number you are seeking. In this paragraph you will find the information concerning the removal of the camshaft.

More information available at haynes.com
Phone: 805-498-6703

Haynes Group Limited
Haynes North America, Inc.

ISBN-10: 0-87288-067-2
ISBN-13: 978-0-87288-067-2

© Haynes North America, Inc. 1990
With permission from Haynes Group Limited

Clymer is a registered trademark of Haynes North America, Inc.

Cover art by Sean Keenan

Disclaimer

There are risks associated with automotive repairs. The ability to make repairs depends on the individual's skill, experience and proper tools. Individuals should act with due care and acknowledge and assume the risk of performing automotive repairs.

The purpose of this manual is to provide comprehensive, useful and accessible automotive repair information, to help you get the best value from your vehicle. However, this manual is not a substitute for a professional certified technician or mechanic.

This repair manual is produced by a third party and is not associated with an individual vehicle manufacturer. If there is any doubt or discrepancy between this manual and the owner's manual or the factory service manual, please refer to the factory service manual or seek assistance from a professional certified technician or mechanic.

Even though we have prepared this manual with extreme care and every attempt is made to ensure that the information in this manual is correct, neither the publisher nor the author can accept responsibility for loss, damage or injury caused by any errors in, or omissions from, the information given.

SHOP MANUAL
JOHN DEERE

IDENTIFICATION

Tractor serial number locations are indicated by white arrows below.

SERIES A, B, G

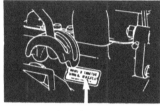
MODEL D

SERIES H

NOTE: Service procedures for M and MT Series tractors begin on page 50.

INDEX (By Starting Paragraph)

SERIES "A"

Serial No., 499000 and up. Bore & Stroke, 5½ x 6¾. Produced in the following versions: Model A, dual wheel tricycle; Model AH, adjustable axle (Hi-Clearance); Model AN, single wheel tricycle; Model ANH, single wheel tricycle (Hi-Clearance); Model AW, adjustable axle; Model AWH, adjustable axle (Hi-Clearance).

SERIES "B"

Serial No., 96000-200999. Bore & Stroke, 4½ x 5½. Serial No., 201000 and up. Bore & Stroke, 4 11/16 x 5½. Produced in the following versions: Model B, dual wheel tricycle; Model BN, single wheel tricycle; Model BNH, single wheel tricycle (Hi-Clearance); Model BW, adjustable axle; Model BWH, adjustable axle (Hi-Clearance).

MODEL "D"

Serial No., 143800 and up. Bore & Stroke, 6¾ x 7. Produced in a non-adjustable axle version only.

SERIES "G"

Serial No., 13000 and up. Bore & Stroke, 6⅛ x 7. Produced in the following versions: Model G, dual wheel tricycle; Model GH, adjustable axle (Hi-Clearance); Model GN, single wheel tricycle; Model GW, adjustable axle; Model GM, dual wheel tricycle.

SERIES "H"

Serial No., 1000 and up. Bore & Stroke, 3 9/16 x 5. Produced in the following versions: Model H, dual wheel tricycle; Model HN, single wheel tricycle; Model HNH, single wheel tricycle (Hi-Clearance); Model HWH, adjustable axle (Hi-Clearance).

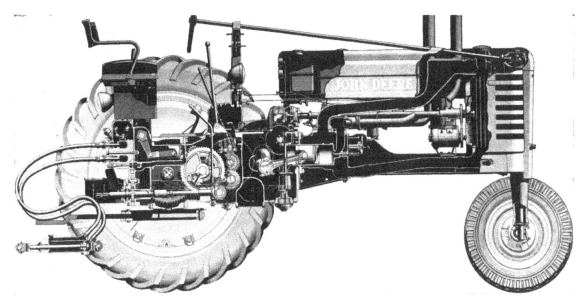

Late production series A. The starter is mounted in the crankcase and the engine is equipped with a crankcase vent pump.

Late production series B. Notice the "Powr-Trol" valve box installation. The engine is equipped with a Wico battery ignition distributor.

Styled series G tractor. The engine is equipped with a Delco-Remy battery ignition unit. The "Hi-Crop" versions are chain driven.

(For Sectional View of Model D, Refer to Page 49)

Sectional view of series H tractor. The belt pulley rotates on the camshaft at one-half engine speed.

CONDENSED SERVICE DATA

Item	A (499000-583999)	A (584000 And Up)	A Gas	B (96000-200999)	B (201000 And Up)	B Gas	D (143800 And Up)	G (13000 And Up)	H (1000 And Up)
ALL-FUEL OR GASOLINE	A-F	A-F	Gas	A-F	A-F	Gas	A-F	A-F	A-F

GENERAL

Item	A (499000-583999)	A (584000 And Up)	A Gas	B (96000-200999)	B (201000 And Up)	B Gas	D	G	H
Torque Recommendations	See End of Shop Manual								
Engine Make	Own	Own	Own	Own	Own	Own	Own	Own	Own
Engine Model	A	A	A	B	B	B	D	G	H
Cylinders	2	2	2	2	2	2	2	2	2
Bore—Inches	5-1/2	5-1/2	5-1/2	4-1/2	4-11/16	4-11/16	6-3/4	6-1/8	3-9/16
Stroke—Inches	6 3/4	6 3/4	6 3/4	5 1/2	5 1/2	5 1/2	7	7	5
Displacement—Cubic Inches	321.2	321.2	321.1	174.9	190.4	190.4	501	412.5	99.68
Compression Ratio (All Fuel)	4.45:1	4.45:1		4.71:1	4.65:1		3.91:1	4.20:1	4.75:1
Compression Ratio (Gasoline)			5.60:1			5.87:1			
Compression Pressure at Cranking Speed	75	75	110	70	70	110	60	65	80
Pistons Removed From:	Front	Front	Front	Front	Front	Front	Front	Front	Front
Main bearings Adjustable?	Yes	Some	Some	Yes	Yes	Yes	Yes	Yes	No
Not Adjustable After Serial		694827	694827						
Rod Bearings Adjustable?	Yes	Yes	Yes	Yes	Yes	Yes	Yes	Yes	Some
Cylinder Sleeves	None	None	None	None	None	None	None	None	None
Forward Speeds	6	6	6	6	6	6	3	6	3
Main Bearings, Number of	2	2	2	2	2	2	2	2	2
Generator Make	D-R	D-R	D-R	D-R	D-R	D-R	D-R	D-R	D-R
Starter Make	D-R	D-R	D-R	D-R	D-R	D-R	D-R	D-R	D-R

TUNE-UP

Item	A (499000-583999)	A (584000 And Up)	A Gas	B (96000-200999)	B (201000 And Up)	B Gas	D	G	H
Valve Tappet Gap—Inlet (Hot)	0.020	0.020	0.020	0.020	0.020	0.020	0.030	0.020	0.015
Valve Tappet Gap—Exhaust (Hot)	0.020	0.020	0.020	0.020	0.020	0.020	0.030	0.020	0.015
Inlet Valve Face Angle	30	30	30	30	29 3/4	29 3/4	30	30	30
Inlet Valve Seat Angle	30	30	30	30	30	30	30	30	30
Exhaust Valve Face Angle	45	45	45	45	44 1/2	44 1/2	30	45	45
Exhaust Valve Seat Angle	45	45	45	45	45	45	30	45	45
Ignition Distributor Make		D-R*	D-R*	Wico*	Wico*		None	D-R*	None
Ignition Distributor Model		111158	111158	XB	XB			1111558	
Ignition Magneto Make	Wico	Wico	Wico	Wico	Wico	Wico	Wico	Wico	Wico
Ignition Magneto Model	X	X	X	X	X	X	X	X	X
Breaker Gap—Distributor		0.021	0.021	0.015	0.015			0.021	
Breaker Gap—Magneto	0.015	0.015	0.015	0.015	0.015	0.015	0.015	0.015	0.015
Distributor Timing—Retard		TC	TC	TC	TC			TC	
Distributor Timing—Full Advance		26°B	26°B	25°B	25°B			26°B	
Magneto Impulse Trip Point	TC	TC	TC	TC	TC	TC	TC	TC	TC
Magneto Lag Angle	25°-35°	25°	25°	25°-35°	25°	25°	35°	25°-30°	25°-35°
Magneto Running Timing	30°B	25°B	25°B	30°B	25°B	25°B	35°B	25°-30°B	30°B
Flywheel Mark Indicating:									
Magneto Impulse Trips	When dot or "LH IMPULSE" aligns with index mark on cover or case								
Distributor Retard Timing	When dot or "LH IMPULSE" aligns with index mark on cover or case								
Spark Plug Make	Champion	Champion	Champion	Champion	Champion	Champion	Champion	Champion	Champion
Model For Gasoline			6			6			
Model for Low Octane	8 Com	8 Com		8 Com	8 Com		2 Com L	8 Com	8 Com
Electrode Gap	0.030	0.030	0.030	0.030	0.030	0.030	0.030	0.030	0.030
Carburetor Make	M-S	M-S	M-S	M-S	M-S	M-S	M-S	M-S	M-S
Model Gasoline			DLTX-71			DLTX-67			
Model All-Fuel	DLTX-24	DLTX-72		DLTX-34	DLTX-73		DLTX-63	DLTX-51	DLTX-46
Float Setting	1/2	3/8	3/8	1/2	3/8	3/8	1/2	1/2	27/32
Engine Rated RPM	975	975	975	1150	1250	1250	900	975	1400
Engine High Idle RPM	1080	1080	1080	1340	1370	1370	980	1115	1540
P.T.O. RPM	545	545	545	554	540	540	525	530	545

SIZES—CAPACITIES—CLEARANCES

(Clearances in thousandths)

Item	A (499000-583999)	A (584000 And Up)	A Gas	B (96000-200999)	B (201000 And Up)	B Gas	D	G	H
Crankshaft Journal Diameter	2.749	2.749	2.749	2.249	2.249	2.249	2.999	2.999	2.0615
Crankpin Diameter	2.999	2.999	2.999	2.7495	2.7495	2.7495	3.4995	3.3745	2.0615
Piston Pin Diameter	1.7495	1.7495	1.7495	1.4165	1.4165	1.4165	1.7495	1.7495	1.1045
Valve Stem Diameter	0.4965	0.4965	0.4965	0.434	0.434	0.434	0.621	0.5585	0.372
Main Brgs., Run. Clearance (Adjustable)	2-6	2-5	2-5	2-6	2-5	2-5	2-6	2-6	2-4
(Sleeve Type)		4-6	4-6						
Rod Brgs. Diam. Clearance	2-4	2-4	2-4	2-4	2-4	2-4	2-4	2-4	1-3
Piston Skirt Clearance	5-8	5-8	5-8	4-7	5-9	5-9	6-10	6-9	3-6
Crankshaft End Play	5-10	5-10	5-10	5-10	5-10	5-10	5-10	5-10	5-10
Cooling System—Gallons	9 1/4	8 1/2	8 1/2	6	7	7	14	13	5 1/2
Crankcase Oil—Quarts	9 1/4	11	11	7 1/2	7	7	13	11	4 1/2
Trans. & Diff.—Quarts	32	27	27	18	18	18	28	28	12
Final Drive, Each—Quarts (AH, GH)	1 3/4	1 3/4	1 3/4					1 3/4	

*Battery ignition adopted as standard equipment in June, 1950. Field installation of the battery ignition units can be made on earlier models.

FRONT SYSTEM—TRICYCLE TYPE

VERTICAL SPINDLE AND/OR FORK
Models A-AN-ANH-B-BN-BNH-G-GM-GN-H-HN-HNH

1. **R & R AND OVERHAUL.** To remove either the wheel fork and vertical spindle assembly on the single wheel version or the vertical spindle and knuckle or "Roll-O-Matic" assembly on the dual wheel version, support front of tractor in a hoist, remove wheel (or wheels) and proceed as follows: Remove medallion, pedestal cover and the nut retaining steering sector to vertical spindle (3—Figs. JD1, 2 & 4). If fork is detachable from the spindle as in Fig. JD1, it may be advisable to detach same at this time. Drive the vertical spindle down until spindle splines are free from the steering sector. Raise front of tractor and withdraw spindle assembly from below.

Inspect all parts and renew any which are excessively worn. The vertical spindle on models B, BN and BNH should have a free fit in lower bushing (16).

On model B, make certain that tangs on lower bushing engage grooves of lower thrust washer (15). On all except series B, make certain that the lower ball thrust bearing is installed with side marked "THRUST" facing down. Refer to paragraphs 18-21 for steering gear unit adjustments.

PEDESTAL
Models A-AN-ANH-B-BN-BNH-G-GM-GN-H-HN-HNH

2. **REMOVE AND REINSTALL.** To remove the pedestal, support front of tractor and remove steering wheel, grilles and screen. Remove steering shaft bearing housing and cover from top of pedestal; then, withdraw steering shaft from front of pedestal. On G, GN and GM models, remove radiator. Remove front wheel (or wheels). Remove nuts retaining pedestal to frame and remove pedestal and vertical spindle assembly. Remove the nut retaining the sector gear to the vertical spindle. Drive the spindle out of sector gear and pedestal. The procedure for further disassembly is evident after an examination of the unit.

Inspect all parts and renew any which are excessively worn. The vertical spindle on models B, BN and BNH should have a free fit in lower bushing (16—Figs. JD 1&2).

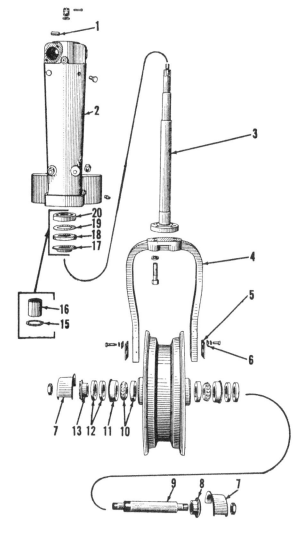

Fig. JD1—Exploded view of model AN steering spindle, wheel fork and associated parts. Models ANH and GN are similar. Models BN and BNH are similarly constructed except that thrust washer (15) and bushing (16), which are shown in the box, are used instead of retainer (17), cork washer (18), washer (19) and bearing (20). Models HN and HNH are similar except that the wheel fork and vertical spindle is an integral unit.

　　1. Sector stop pin
　　2. Pedestal
　　3. Vertical spindle
　　4. Wheel fork
　　5. Axle lock plate
　　6. Axle nut lock plate
　　7. Dust shield
　　8. Bearing spacer
　　9. Front axle
　10. Bearing
　11. Retainer
　12. Felt washer
　13. Bearing adjusting nut

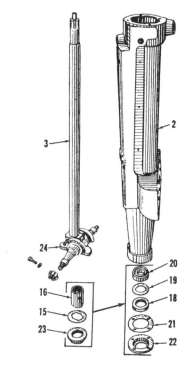

Fig. JD2—Exploded view of model A pedestal and vertical spindle assembly. Models G and GM are similar. Model B is similarly constructed except that thrust washer (15), bushing (16) and pivot plate (23) are used in place of cork washer (18), washer (19), bearing (20), gasket (21) and cap (22). Model H is similar except that a retainer is used instead of gasket (21) and cap (22).

　2. Pedestal　　　　　　　3. Vertical spindle
　　　　　　　24. Knuckle

On model B, make certain that tangs on lower bushing engage grooves of lower thrust washer (15). On all except series B, make certain that the lower ball thrust bearing is installed with side marked "THRUST" facing down. Refer to paragraphs 18-21 for steering gear unit adjustments.

"ROLL-O-MATIC"

Models A-B-G

3. **OVERHAUL.** To disassemble the "Roll-O-Matic" unit, remove cap (29 —Fig. JD4), thrust washer (31) and pull knuckle and gear from housing. Each knuckle is fitted with two steel-backed, bronze-lined bushings which can be pressed from the knuckles. Install knuckle bushings so that open end of oil groove is toward inside, and make certain that there is a $\frac{1}{32}$-$\frac{1}{16}$-inch gap between the bushings. Pack the "Roll-O-Matic" housing with wheel bearing grease and install a new felt washer (25). When meshing gear teeth, make certain that punch marks are in register.

FRONT SYSTEM—AXLE TYPE

STEERING KNUCKLES
Models AH-AW-AWH-BW-BWH-GH-GW-HWH

4. **R&R AND OVERHAUL.** Method of removal is self evident. Spindle bushings (34 and 55—Figs. JD5 and JD6) can be driven from the axle knee and new bushings can be pressed in. Bushings are pre-sized to provide 0.009-0.012 clearance. On model HWH, make certain that long bushing (55—Fig. JD6) is installed at lower end of knee.

When reassembling make certain that the dowel pins engage thrust washers (47). Vary the number of shim washers to provide not more than 0.036 vertical end play for the spindle.

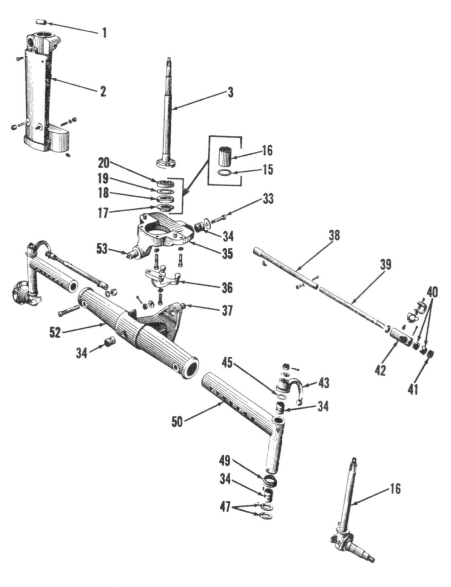

Fig. JD5—Exploded view of model AW front axle and related parts. Models AH, AWH, GH and GW are similar. Models BW and BWH are similarly constructed except that bushing (16) and thrust washer (15) are used instead of retainer (17), cork washer (18), washer (19) and bearing (20).

1. Sector stop pin	36. Center steering arm
2. Pedestal	37. Rear pivot pin
3. Vertical spindle	38. Tie rod (drag link) sleeve
16. Knuckle and spindle	39. Tie rod (drag link)
33. Bolt	40. Ball stud bearings
34. Bushing	41. Screw plug
35. Axle pivot bracket	42. Tie rod (drag link) end

43. Steering arm
45. Shim washer
47. Thrust washers
49. Dust shield
50. Axle knee
52. Axle center member
53. Front pivot pin

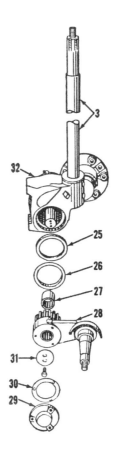

Fig. JD4—Partially exploded view of Deere models A, B and G "Roll-O-Matic" unit.

25. Felt washer	29. Cap
26. Felt retainer	30. Gasket
27. Bushing (4 used)	31. Thrust washer
28. Knuckle	32. Knuckle housing

Model D

5. **R&R AND OVERHAUL.** Remove wheel and hub assembly, disconnect steering arm from knuckle, remove nut from taper bolt (74—Fig. JD8) and drive bolt out of axle main member. Remove knuckle caps (69) and pins (72). Knuckle bushings (70) are pre-sized and if not distorted during installation, will require no final sizing.

TIE RODS AND DRAG LINKS
Models AH-AW-AWH-BW-BWH-D-GH-GW-HWH

7. **ADJUSTMENT.** An adjustable ball-socket is fitted to both ends of each tie rod, and on model D, an adjustable ball-socket is fitted to both ends of the drag link. Tie-rod and drag link ends should be adjusted so they have no end play, yet do not bind.

With steering gear centered and wheels pointing straight ahead, adjust tie rod length to provide a recommended toe-in of ⅛-³⁄₁₆ inch.

AXLE PIVOT PINS & BUSHINGS
Models AH-AW-AWH-BW-BWH-GH-GW

8. To renew the front axle pivot pin bushings, support front of tractor and remove the bolt (33—Fig. JD5) which retains pivot pin (37) in the front axle pivot bracket. On GH models so equipped, disconnect radius rod from axle knees. Disconnect tie rods from center steering arm. Slide axle assembly forward until axle is free from pivots. Pivot pin bushings which are interchangeable with the knuckle and spindle bushings are pre-sized, and if not distorted during installation, will require no final sizing.

If axle pivot pins are to be renewed, remove the center steering arm and pivot bracket and press the pivot pins out of axle and pivot bracket. A mild application of heat will facilitate removal and installation of pins.

Model HWH

9. To renew pivot pin bushings, support front of tractor and remove rear pivot pin (37—Fig. JD6). Slide axle forward and off front pivot pin (53). Then, remove pivot bracket (35) from tractor. Bushings for bracket and axle are pre-sized, and if not distorted during installation, will require no final sizing.

If front pivot pin is to be renewed, remove pedestal from tractor and press the pin out of pedestal. A mild application of heat on pedestal will facilitate removal and installation of the front pivot pin.

Model D

11. The unbushed front axle pivots on pin (76—Fig. JD8). The front axle pivot pin should have a clearance of 0.015 (I&T recommended) in the axle. Method of removal of the pivot pin is evident after an examination of the unit and reference to Fig. JD8.

RADIUS ROD
Model D

12. **R&R AND OVERHAUL.** To remove the radius rod, jack up front end of tractor and disconnect drag link from right hand steering arm. Remove axle pivot pin, radius rod pivot stud nut and roll axle and radius rod assembly away from tractor. Remove radius rod to axle retaining nuts and bump radius rod out of axle main member.

The radius rod pivot stud bushing (61) can be removed from pivot block (62) at this time. After new bushing is installed, check to make certain that the radius rod pivot stud has a free fit in the bushing.

Model GH (Some)

13. **R&R AND OVERHAUL.** Removal of the adjustable type radius rod is evident after an examination of the unit and reference to Fig. JD9. Renew pivot (62) and/or pivot bracket (80) if either is excessively worn.

After installing the radius rod—or changing the front wheel tread, make certain that jam nuts (82) are tight.

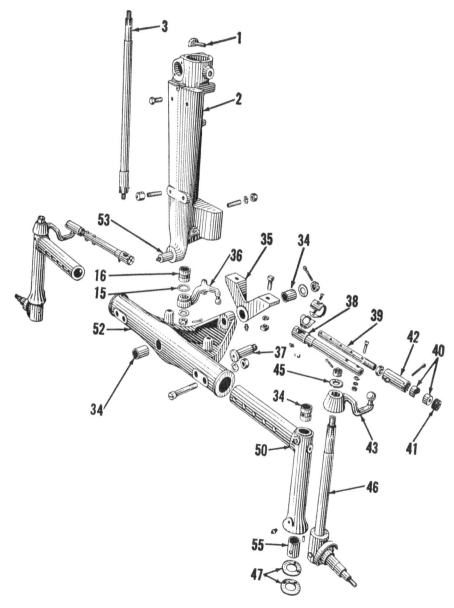

Fig. JD6—Exploded view of model HWH adjustable front axle and associated parts.

1. Sector stop pin	37. Rear pivot pin	45. Shim washer
3. Vertical spindle	38. Tie rod (drag link) tube	46. Knuckle and spindle
15. Thrust washer	39. Tie rod (drag link)	47. Thrust washers
16. Bushing	40. Ball stud bearings	50. Axle knee
34. Bushing	41. Screw plug	52. Axle center member
35. Pivot bracket	42. Tie rod (drag link) end	53. Front pivot pin
36. Center steering arm	43. Steering arm	55. Lower knuckle bushing

VERTICAL SPINDLE
Models AH-AW-AWH-BW-BWH-GH-GW-HWH

15. **R&R AND OVERHAUL.** To remove the vertical spindle, support front of tractor and remove the bolt (33—Fig. JD5) or on model HWH, remove rear pivot pin (37—Fig. JD6). Disconnect center steering arm from vertical spindle and on GH models so equipped, disconnect radius rod from axle knees. Slide axle assembly forward until free from pivot pins. Remove medallion, pedestal top cover and nut which retains sector gear to vertical spindle. Drive the vertical spindle out of sector gear and pedestal.

Inspect all parts and renew any which are excessively worn. The vertical spindle on models BW, BWH, and HWH should have a free fit in lower bushing (16—Fig. JD5 or JD6).

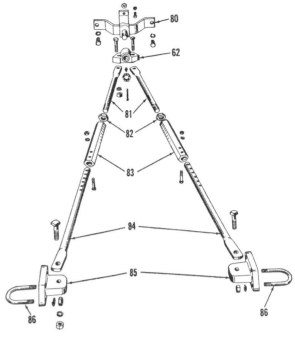

Fig. JD9—Exploded view of radius rod assembly which is used on some GH tractors.

62. Pivot
80. Pivot bracket
81. Rear radius rod
82. Jam nuts
83. Coupling
84. Front radius rod
85. Radius rod bracket
86. U-Bolt

On all except series B, make certain that the lower ball thrust bearing is installed with side marked "THRUST" facing down. Refer to paragraphs 18-21 for steering gear unit adjustments.

PEDESTAL
Models AH-AW-AWH-BW-BWH-GH-GW-HWH

16. **REMOVE AND REINSTALL.** To remove pedestal, support front of tractor and remove steering wheel, grilles and screen. Remove steering shaft bearing housing and top cover from pedestal; then, withdraw steering shaft from front of pedestal. Remove the bolt (33—Fig. JD5) or on model HWH, remove rear pivot pin (37—Fig. JD6). Disconnect center steering arm from vertical spindle and on GH models so equipped, disconnect radius rod from axle knees. Slide axle assembly forward until axle is free from pivot pins.

On models GH and GW, remove radiator. Remove nuts retaining pedestal to frame and remove pedestal and vertical spindle assembly. Remove the nut retaining the sector gear to the vertical spindle and drive the vertical spindle out of sector gear and pedestal. The procedure for further disassembly is evident after an examination of the unit.

The vertical spindle on models BW, BWH and HWH should have a free fit in bushing (16—Figs. JD5 or JD6). On all except series B, make certain that the lower thrust bearing is installed with side marked "THRUST" facing down. Refer to paragraphs 18-21 for steering gear unit adjustments.

Fig. JD8—Exploded view of model D front axle and associated parts. Vertical steering shaft tube (58) and bushing (57) are carried in the tractor main case.

36. Center steering arm	65. Tie rod end	72. Knuckle pin
56. Vertical steering shaft	66. Ball stud bearings	73. Knuckle
59. Felt washer	67. Screw plug	74. Taper bolt
60. Stud	68. L.H. steering arm	75. Axle stop
61. Bushing	69. Dust cap	76. Pivot pin
62. Pivot block	70. Bushing	77. Axle
63. Radius rod	71. Thrust washers	78. R.H. Steering arm
64. Tie rod		79. Drag link

STEERING GEAR

Series A-B-G-H

For the purposes of this discussion the steering gear unit will include the steering shaft and worm, sector gear, and all parts contained in the upper portion of the pedestal. For R&R and overhaul of the vertical spindle, refer to paragraph 1 or 15.

18. **ADJUSTMENT.** Three adjustments are provided on the steering gear unit: (1) steering shaft (worm-shaft) end play, (2) vertical spindle end play and (3) backlash between the worm and sector gears.

19. **STEERING SHAFT (WORM-SHAFT) END PLAY.** Support front of tractor to remove load from steering gear. Remove grille medallion and vary the number or thickness of shim gaskets (88—Fig. JD11) between the pedestal and the steering shaft front bearing housing to remove all end play, yet permit shaft to rotate freely.

20. **VERTICAL SPINDLE END PLAY.** Support front of tractor to remove load from steering gear and remove the grille medallion and pedestal top cover. Remove nut retaining sector gear to vertical spindle, and using a wheel knocker or similar tool, remove sector (92— Fig. JD11). Vary the number of shim washers (95) under flange of eccentric bushing (94) to obtain not more than $\frac{1}{32}$-inch end play.

21. **BACKLASH.** First make certain that the end play of wormshaft and of vertical spindle is properly adjusted as outlined in paragraphs 19 and 20 then remove eccentric bushing lock screw (93—Fig. JD11). Turn the bushing (94) to obtain a backlash of ½-1 inch (measured at rim of steering wheel) between the two gears.

23. **OVERHAUL.** To overhaul the steering gear unit, support front of tractor and remove grille medallion, pedestal top cover and steering wheel. Remove wormshaft from pedestal. Steering shaft bearing cones can be renewed at this time and the front bearing cup can be pulled from the bearing housing. To remove the wormshaft rear bearing cup, it is necessary to remove the pedestal as outlined in paragraph 2 or 16. Remove the nut retaining the sector gear to the vertical spindle, and using a wheel knocker or similar tool, remove the sector. The need

and procedure for further disassembly is evident after an examination of the unit and reference to Fig. JD11. Adjust the unit as outlined in paragraphs 18-21.

Model D

25. **ADJUSTMENT.** Three adjustments are provided on the model D steering gear unit; (1) steering worm-shaft end play, (2) vertical shaft end play and (3) backlash between the worm and the steering gear.

26. **WORMSHAFT END PLAY.** Support front of tractor to remove load from steering gear and remove hood. Loosen the adjusting nut lock bolt and tighten adjusting nut (99—Fig. JD13) until the wormshaft has zero end play, yet turns freely.

27. **VERTICAL SHAFT END PLAY.** Support front of tractor to remove load from steering gear and remove hood. Loosen jam nut (102) and turn adjusting screw (103) **down** until the

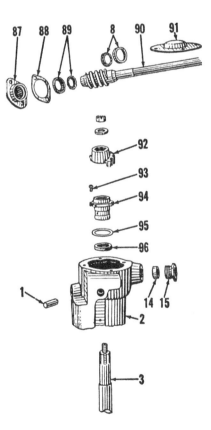

Fig. JD11—Exploded view of series A, B, G & H steering gear.

1. Sector stop pin	91. Pedestal cover
3. Vertical spindle	92. Sector
87. Bearing housing	93. Lock screw
88. Shim gaskets	94. Eccentric bushing
89. Bearing	95. Shim washers
90. Wormshaft	96. Packing

vertical shaft has zero end play, then back the screw off ⅛ turn and tighten the jam nut.

28. **BACKLASH.** First remove hood and either remove dash, or unbolt it and slide the dash far enough up the steering shaft to gain access to the steering gear housing. Remove the cap screws joining the wormshaft housing and the steering gear housing. Separate the two housings and vary the thickness of shim gaskets (108) between the worm housing and gear housing to provide ½-1 inch free movement measured on rim of steering wheel.

NOTE: Make certain that the housing locating dowels are in place before reconnecting the two housings.

Check for binding through entire range of steering wheel travel. If the steering gear binds or has an excessive amount of backlash in any position, it will be necessary to renew the worm and worm wheel or, reposition the gears so as to bring unworn teeth into mesh. To re-position the gears, turn steering wheel to one extreme position and detach steering gear arm (36) from the vertical steering shaft. Then move front wheels to other extreme position and reconnect the steering gear arm. Recheck backlash after re-positioning the steering gears.

29. **R&R AND OVERHAUL.** To remove the steering gear unit, remove hood and steering wheel and disconnect the steering shaft upper bearing from its support. Remove dash and detach steering gear arm (36—Fig. JD13) from the vertical steering shaft. Remove cap screws retaining the steering gear housing to the main case and lift gear unit from tractor.

Vertical shaft lower bushing (57), which is driven into the bottom portion of the tractor main case can be renewed at this time. The bushing is pre-sized, and if not distorted during installation, will require no final sizing.

Procedure for disassembly and reassembly of the gear unit is evident. Renew any parts which are excessively worn, and make certain that thrust washer (107) is installed between the worm wheel (106) and the gear housing. Adjust the unit as outlined in paragraphs 25-28.

30. **STEERING SHAFT TUBE—RE-NEW.** To renew the vertical steering shaft tube, (58—Fig. JD13), first remove the gear unit as in paragraph 29, drive bushing (57) out through bottom of main case and drive the tube out through top of main case. Install new tube from above and using a piloted driver, drive the tube down until lower shoulder of tube bottoms in main case. Install lower bushing (57) and the steering gear unit.

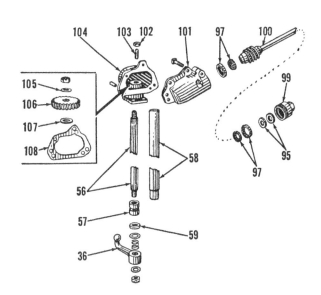

Fig. JD13—Exploded view of model D steering gear unit. The vertical steering shaft tube (58) and bushing (57) are carried in the tractor main case.

36. Center steering arm
56. Vertical steering shaft
59. Felt washer
95. Cork washers
97. Bearing
99. Adjusting nut
100. Worm and steering shaft
101. Worm housing
102. Lock nut
103. Adjusting screw
104. Gear housing
105. Washer
106. Steering gear
107. Thrust washer
108. Shim gaskets

ENGINE AND COMPONENTS

The engine crankcase and tractor main case is an integral unit. A wall in the main case separates the engine crankcase compartment from the transmission and differential compartment.

CYLINDER HEAD—R&R

Series A-B-G

35. To remove the cylinder head, first drain cooling system, remove lower water pipe and on models so equipped, remove the front mounted tool box. Remove exhaust pipe, and on models so equipped, remove generator and mounting bracket assembly. Disconnect fuel line and controls and remove carburetor. On series A584000 and up, B201000 and up, and G, remove air inlet elbow. Remove lines connected to tappet lever cover and remove the tappet lever cover. Turn flywheel until all valves are closed and remove tappet levers assembly and push rods. On series A prior 584000, remove left tappet lever bracket retaining stud to prevent damaging the air cleaner. Remove the cylinder head retaining stud nuts and slide cylinder head forward and off studs. On series B prior to 201000, tip cylinder head down and lift out. On series A prior 584000, tip head backwards so that springs are up and lift head out. On series A584000 and up, series B201000 and up and series G, turn left side of head toward front (valve springs toward right) and roll head out over right side of frame.

NOTE: On series G tractors it may be necessary to remove the top four cylinder head retaining studs to provide removal clearance.

36. Soak the cylinder head gasket in engine oil for five seconds and allow to drain for one minute before installation. Install cylinder head gasket with smooth side toward cylinder block and use new lead washers under the cylinder head retaining stud nuts. If cylinder head or tappet lever retaining studs were removed, apply white lead or equivalent sealing compound to the threads before installing studs. Tighten cylinder head retaining stud nuts from center outward and to a torque of 125 ft.-lbs. for series A; 96 ft.-lbs. for series B; and 208 ft.-lbs. for series G. On all heat treated studs (dark colored studs) tighten the stud nuts to a torque of 150 ft.-lbs. for series A; 104 ft.-lbs. for series B. Adjust tappet gap (hot) to 0.020. Before installing tappet lever cover, make certain that tappet lever oiler tube (or trough) is clean and properly installed.

Model D

39. To remove the cylinder head, remove radiator as outlined in paragraph 151 and proceed as follows: Remove fan and driving discs and generator drive pulley. Disconnect fuel line, controls, water feed pipe and remove carburetor. Remove tappet cover inspection plate, turn flywheel until all valves are closed and remove tappet cover. Withdraw push rods, remove cylinder head retaining stud nuts and slide cylinder head forward and off studs.

40. Soak the cylinder head gasket in engine oil for five seconds and allow to drain for one minute before installation. Install gasket with smooth side toward block and use new lead washers under the nuts. Tighten stud nuts from center outward and to a torque of 208 ft.-lbs. Adjust tappet gap to 0.030 (hot). *NOTE: 0.030 equals ½ turn of the tappet adjusting screw.*

Before installing tappet lever cover, make certain that oil lines are clean and free from obstructions. Tighten fan shaft nut to a torque of 10 ft.-lbs.

Series H

41. To remove cylinder head, remove generator and battery (if tractor is so equipped) and drain cooling system. Remove exhaust pipe and lower water pipe. Disconnect fuel lines and controls and remove carburetor and air cleaner cup. Remove tappet cover and oil line connected to cylinder head retaining stud. Loosen tappet adjusting screws so as to remove tension from valve springs and remove tappet lever shaft bearing caps from sides of head. Withdraw tappet lever shaft from side of head and remove levers and push rods from front of head. Remove cylinder head retaining stud nuts and slide head off studs.

42. Soak the cylinder head gasket in engine oil for five seconds and allow to drain for one minute before installation. Install gasket with smooth side toward block and use new lead washers under the nuts. Tighten stud nuts from center outward and to a torque of 96 foot pounds for all except the ¾ inch oil lead stud which should be torqued to 180 ft.-lbs. Remove crankcase cover when installing push rods so that rods can be guided into cam followers. Adjust tappet gap to 0.015 (hot).

PUSH ROD SLEEVES

Series B

Push rod sleeves (S—Fig. JD20) should be examined whenever cylinder head is removed. Renew any sleeve that is corroded, deteriorated or shows signs of coolant leakage. Sleeves are sometimes damaged by manifold studs that have been installed with wrong end of stud in cylinder head.

45. **RENEW.** To remove push rod sleeves when cylinder head is off, use a drift or special O.T.C. puller, and remove sleeves through front of head. Sleeves can be removed when cylinder head is on by removing the tappet levers and push rods and using a special O.T.C. puller or its equivalent.

Install sleeves from front of head with the smaller outside diameter toward crankcase.

NOTE: Rear O.D. of sleeve is approximately 0.002 smaller than front. Use white lead or equivalent sealing compound on ends of sleeve to seal the sleeve against coolant leakage and facilitate installation.

VALVES AND SEATS

Series A-B-G-H-Model D

46. Valves are provided with split-cone type keepers. Set tappet gap (hot) to the values specified below.

Series A, B, G..................0.020
Series H 0.015
Model D 0.030

47. Intake and exhaust valves are not interchangeable and seat directly in cylinder head. Check the valves and seats against the values listed below. Seats can be narrowed, using 20 and 70 degree stones.

Valve stem diameter
Series A..............0.4960-0.4970
Series B..............0.4335-0.4345
Series G..............0.558-0.559
Series H..............0.3715-0.3725
Model D..............0.620-0.622

Valve stem clearance in guides
Series A and B........0.004-0.0065
Series G.............0.004-0.0075
Model D.............0.004-0.008
Series H.............0.0035-0.006

Valve face angle (Intake)
Series A, B (prior 201000),
  G and H.....................30°
Series B (after 200999).......29¾°
Model D.......................30°
Valve seat angle (Intake)........30°
Valve face angle (Exhaust)
Series A, B (prior 201000),
  G and H.....................45°
Series B (after 200999).......44½°
Model D.......................30°
Valve seat angle (Exhaust)
Series A, B, G, H.............45°
Model D.......................30°

Valve seat width
Series A, B, H............1/8-inch
Series G...............11/64-inch
Model D.................5/32-inch

49. When servicing the valves, make certain that the tappet lever lubrication system is functioning properly. On engines equipped with an oil trough in the tappet lever cover, clean the trough and make certain that drip holes are open. On models equipped with a pressure fed drip tube, clean the tube and make certain that holes in tube are so positioned that oil is directed on the exhaust valves only.

VALVE GUIDES AND SPRINGS

Series A-B-G-H-Model D

50. Intake and exhaust valve guides are interchangeable and should be pressed into the cylinder head to the dimensions listed below. Ream the guides after installation (if necessary) to the dimensions listed below.

Distance from end of guide to gasket face of cylinder head.
Series A (prior 584000)...2⅛ inches
Series A (after 583999)...1⅞ inches
Series B (intake).......1$\frac{31}{32}$ inches
Series B (exhaust)......1$\frac{23}{32}$ inches
Series G.................2¼ inches
Model D..................1⅝ inches
Series H................1$\frac{11}{16}$ inches

Ream guides I.D. to:
Series A.............0.5010-0.5025
Series B.............0.4385-0.4400
Series G.............0.5635-0.5650
Series H.............0.3760-0.3775
Model D.............0.6260-0.6280

51. Renew valve springs that are rusted, distorted or do not meet the pressure test specifications which follow.

Valve spring length (inches) @ lbs. pressure:
Series A...............2¾ @ 36-44
Series B...............2$\frac{13}{16}$ @ 35-39
Series G...............3⅜ @ 56-58
Model D...............3⅜ @ 58-72
Series H...............2⅝ @ 62-68

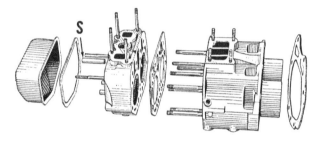

Fig. JD20—Exploded view of series B cylinder block, cylinder head and associated parts. Series A and G are similar. Tappet rod (push rod) sleeves (S) are used on series B only.

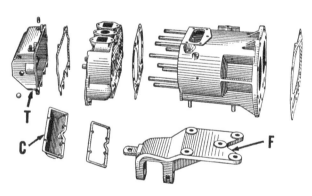

Fig. JD21—Exploded view of model D cylinder block and associated parts. The valve lever (rocker arm) shaft is retained in tappet case (T). Tappets can be adjusted after removing inspection cover (C). The front end support (F) is bolted to bottom face of cylinder block.

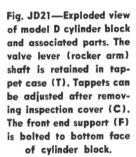

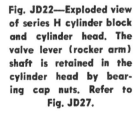

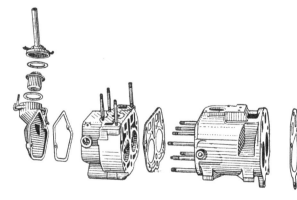

Fig. JD22—Exploded view of series H cylinder block and cylinder head. The valve lever (rocker arm) shaft is retained in the cylinder head by bearing cap nuts. Refer to Fig. JD27.

VALVE TAPPET LEVERS (ROCKER ARMS)

Series A-B-G

53. **R&R AND OVERHAUL.** To remove the valve tappet levers and shaft assembly, first remove the front mounted tool box if tractor is so equipped; then, remove the valve tappet lever cover.

NOTE: On some tractors, the tappet lever cover ventilator pipe must be removed and on others, the tappet oiler line must be disconnected before the cover can be removed. Turn flywheel until all valves are closed, then remove tappet levers assembly. Disassembly of the unit is self-evident.

Excessive wear of any of the component parts of the tappet lever assembly is corrected by renewing the parts. Adjust tappets to 0.020 (hot). Make certain that tappet levers lubrication system is functioning properly; refer to paragraph 49.

Model D

55. **R&R AND OVERHAUL.** To remove the valve tappet levers and shaft, remove radiator as outlined in paragraph 151 and proceed as follows: Remove tappet lever inspection plate, turn flywheel until all valves are closed and remove tappet cover from cylinder head. Remove set screws (5—Fig. JD26) and expansion plugs from ends of tappet lever shaft. Withdraw shaft and levers from cover.

Excessive wear in the tappet lever assembly is corrected by renewal of the component parts. Apply a light coat of sealing compound to the expansion plugs before installing the plugs in the tappet lever cover. Adjust tappets to 0.030 (hot), which equals ½ turn of the adjusting screw. Make certain that tappet levers lubrication system is functioning properly; refer to paragraph 49.

Series H

56. **R&R AND OVERHAUL.** To remove valve tappet levers and shaft, remove tappet lever cover, loosen the tappet adjusting screws to remove any valve spring pressure from the tappet lever shaft and remove tappet lever shaft bearing cap nuts (5—Fig. JD27). Withdraw the tappet lever shaft from side of cylinder head and levers from the front.

NOTE: Intake and exhaust tappet levers are not interchangeable and the right exhaust tappet lever is not interchangeable with the left.

Excessive wear in the tappet lever assembly is corrected by renewal of the component parts. Adjust tappets to 0.015 (hot). Make certain that tappet levers lubrication system is functioning properly; refer to paragraph 49.

VALVE TIMING

Series A-B-G-H-Model D

60. Valves are properly timed when camshaft gear and crankshaft gear are meshed so that timing marks are in register.

To check valve timing when engine is assembled, first adjust tappets (hot) to 0.020 for series A, B and G; 0.015 for series H; and 0.030 for model D; then, check the position of the flywheel on the crankshaft as outlined in paragraph 110.

If the flywheel has been installed incorrectly, remove flywheel and reinstall in the correct position. Turn flywheel in normal direction of rotation until exhaust valve of left cylinder is just beginning to open (0.000 tappet gap). At this time, the flywheel mark "L.H. EXHAUST OPEN" should be in register or within 1 inch (for all except Series H) of index mark on gear case or sliding gear shaft cover as shown in Fig. JD30. On series H, the mark should be within ½ inch. If timing marks are not as specified, it will be necessary to retime the valves as outlined in one of the appropriate paragraphs which follow.

61. On A and G series tractors, loosen governor case retaining cap screws and raise governor until governor gear and camshaft gear are out of mesh. Remove flywheel and the camshaft left bearing. Lift end of camshaft up so that camshaft gear and crankshaft gear are out of mesh; then, turn crankshaft until valve timing is as specified. Recheck tappet adjustment and reset ignition timing as in paragraph 162 or 163.

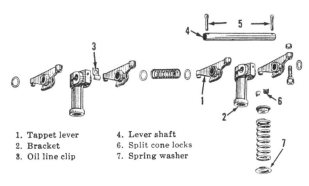

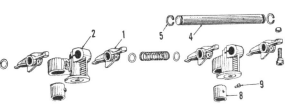

1. Tappet lever
2. Bracket
3. Oil line clip
4. Lever shaft
6. Split cone locks
7. Spring washer

Fig. JD24—Exploded view of series B valve levers assembly. Series A is similarly constructed except that tappet levers are retained by snap rings instead of cotter pins (5) also, spring washers (7) are not used.

Fig. JD25—Exploded view of series G valve levers assembly. Exhaust valve pushrod sleeves (8) are prevented from turning in brackets (2) by sleeve pins (9).
1. Valve lever
4. Lever shaft
5. Snap ring

Fig. JD26—Exploded view of model D valve levers assembly. Lever shaft (4) is retained in tappet cover by set screw (5). Adjusting screws (11) are locked by screws (12).

Fig. JD27—Exploded view of series H valve levers assembly. Lever shaft (4) is retained in the cylinder head by bearing cap nuts (5).

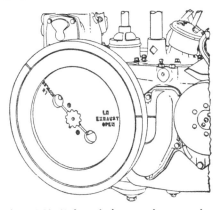

Fig. JD30—Valve timing mark on series B flywheel aligned with mark on cover when No. 1 cylinder (left cylinder) exhaust valve opens. Series A, G and H and model D are similar.

62. On B series tractors, remove clutch and belt pulley as outlined in paragraph 177. Remove crankcase cover, reduction gear cover and camshaft right bearing. Loosen governor case retaining cap screws and remove camshaft left bearing. Lift end of camshaft up so that camshaft gear and crankshaft gear are out of mesh; then turn crankshaft until valve timing is as specified. Recheck tappet adjustment and ignition timing.

63. On model D tractors, loosen camshaft right bearing and remove camshaft left bearing. Push end of camshaft down so that camshaft gear and crankshaft gear are out of mesh; then, turn crankshaft until valve timing is as specified. Recheck tappet adjustment.

64. On H series tractors, loosen the governor case retaining cap screws and remove camshaft left bearing housing (or cover on early models) and remove the left bearing. Remove belt pulley and clutch assembly as outlined in paragraph 180, then remove the reduction gear cover and loosen the camshaft right bearing housing. Lift end of camshaft up so that camshaft gear and crankshaft gear are out of mesh; then, turn crankshaft until valve timing is as specified. Recheck tappet adjustment and ignition timing.

TIMING GEARS
Series A-B-G-H-Model D

70. **CAMSHAFT GEAR.** To remove camshaft gear, it is first necessary to remove camshaft as outlined in paragraph 75, 78 or 79.

71. **CRANKSHAFT GEAR.** To remove the crankshaft gear, it is first necessary to remove the crankshaft as outlined in paragraph 103, 106 or 108. Heat gear in boiling water to facilitate installation.

72. **GOVERNOR GEAR.** To remove the governor gear, it is first necessary to remove the governor as outlined in paragraph 142, 145 or 148.

CAMSHAFT AND BEARINGS
Series A-B-G

75. **REMOVE AND REINSTALL.** To remove camshaft and bearings, first remove governor assembly as outlined in paragraph 142 and belt pulley and clutch assembly as outlined in paragraph 177; then proceed as follows: Remove crankcase cover, oil pump cover and disengage pump drive shaft from the drive coupling. Remove right brake assembly and reduction gear cover. On series A, B & G, remove the flywheel. Remove tappet lever cover and loosen tappet adjust-

ing screws. Mark the camshaft gear with respect to the camshaft to assure correct assembly. Remove camshaft left bearing. On series B tractors, disconnect oil line and nipple from camshaft right bearing and remove the right bearing. Working inside of crankcase, disconnect oil lines which connect to top of main case. On all models remove cap screws retaining cam follower bracket to main case and cap screws and/or nut attaching camshaft gear to camshaft. Push camshaft out of the camshaft gear, remove gear and withdraw camshaft.

On A and G series tractors, the camshaft end play is controlled by thrust spring (5—Fig. JD33) which is located behind the camshaft right bearing cup. On B series tractors, the

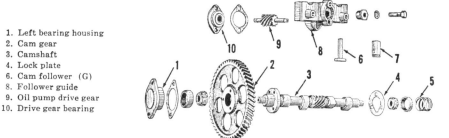

1. Left bearing housing
2. Cam gear
3. Camshaft
4. Lock plate
6. Cam follower (G)
8. Follower guide
9. Oil pump drive gear
10. Drive gear bearing

Fig. JD33—Exploded view of series G camshaft and associated parts. Series A is similar except cam followers (7) are of the barrel type. Camshaft end play is controlled by thrust spring (5).

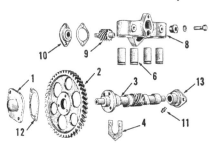

Fig. JD34—Exploded view of series B camshaft, bearings and cam followers. Camshaft end play is controlled by shim gaskets (12) under the left bearing.

1. Left bearing	8. Follower guide
2. Camshaft gear	9. Oil pump drive gear
3. Camshaft	10. Drive gear bearing
4. Lock plate	11. Oil line nipple
6. Cam follower	13. Right bearing

recommended camshaft end play of 1/64-1/32 inch is obtained by varying the number of shims (12—Fig. JD34).

When reinstalling the shaft, align the previously affixed marks on the camshaft gear and camshaft and mesh camshaft gear with crankshaft gear so that timing marks are in register. Mesh governor drive gear with camshaft gear so that groove in magneto or distributor drive coupling flange is horizontal when depressed dot on flywheel hub or flywheel mark "L.H. IMPULSE" is in register with mark on cover or gear case. Adjust tappets to 0.020 hot and check ignition timing.

Model D

78. **REMOVE AND REINSTALL.** To remove camshaft, remove crankcase cover, tappet inspection plate and loosen the tappet adjusting screws. Loosen bolts clamping the camshaft gear to camshaft. Remove belt pulley and clutch assembly as outlined in paragraph 177, remove the reduction gear cover and remove the camshaft right bearing. Remove camshaft left bearing, pull camshaft out of camshaft gear and withdraw shaft through right bearing opening in crankcase.

Recommended camshaft end play of 1/64-1/32 inch is obtained by varying the number of shim gaskets (12—Fig. JD35).

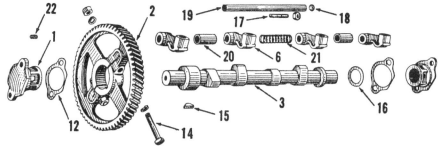

Fig. JD35—Exploded view of model D camshaft, bearings and cam followers. Camshaft end play is controlled by shim gaskets (12).

1. Camshaft bearing	14. Clamp bolt	18. Expansion plug
2. Camshaft gear	15. Woodruff key	19. Cam follower shaft
3. Camshaft	16. Spacer washer	20. Spacer
6. Cam follower	17. Set screw	21. Spacer spring
12. Shim gasket		22. Pipe plug (in crankcase)

When reinstalling be sure that Woodruff key (15) is in position in camshaft and gear and that steel spacer washer (16) is installed on right end of camshaft. Mesh camshaft gear with crankshaft gear so that timing marks are in register and mesh governor gear with camshaft gear so that groove in magneto coupling flange is horizontal when flywheel mark "L.H. IMPULSE" aligns with mark on cover or gear case. Adjust tappets to 0.030 hot and check ignition timing.

Series H

79. **REMOVE AND REINSTALL.** To remove camshaft, remove governor assembly as outlined in paragraph 148 and belt pulley and clutch assembly as outlined in paragraph 180. Remove crankcase cover, tappet lever cover and loosen the tappet adjusting screws. Remove reduction gear cover and camshaft right bearing housing, being careful not to damage oil seal (26— Fig. JD37). Mark camshaft left bearing housing with respect to the crankcase so that housing can be reinstalled in same position. Remove camshaft left bearing, detach gear from camshaft and withdraw shaft through right camshaft bearing opening in crankcase. Pulley journal diameter is 1.499-1.500.

NOTE: Camshaft end play is non-adjustable.

Reinstall camshaft with "V" mark on camshaft gear in register with "V" mark on camshaft. Mesh governor gear with camshaft gear so that groove in magneto coupling flange is horizontal when flywheel mark "L.H. IMPULSE"

is in register with mark on cover or gear case. Adjust tappets to 0.015 hot and check ignition timing.

NOTE: Gear backlash is controlled by the relative position of the eccentrically machined camshaft left bearing housing with respect to the main case. To reduce the gear backlash, slide the left bearing off studs, turn the housing 1/4 turn to the right and reinstall the housing.

CAM FOLLOWERS

Series A-B-G

81. **REMOVE AND REINSTALL.** To remove the cam followers and guide assembly, remove camshaft as outlined in paragraph 75. Remove cap screws retaining follower guide (8— Figs. JD33 or 34) to main case and remove followers and guide.

Reinstall cam followers unit and adjust tappets to 0.020 (hot).

Model D

82. **REMOVE AND REINSTALL.** To remove the cam followers, first remove the radiator as outlined in paragraph 151, then remove the tappet lever cover and push rods. Remove crankcase cover and set screw (17— Fig. JD35) which retains cam follower shaft (19) in right side of crankcase. Remove pipe plug (22) from left side of crankcase, drive the cam follower shaft and expansion plug out through right side of crankcase and withdraw cam followers through crankcase cover opening.

Reinstall in reverse order and seal expansion plug (18) with white lead or equivalent sealing compound. Ad-

just tappets to 0.030 (hot), which equals 1/2 turn of the adjusting screw.

Series H

83. **REMOVE AND REINSTALL.** To remove cam followers, first remove generator and battery if tractor is so equipped. Remove tappet lever cover, loosen tappet adjusting screws to release any valve spring load from the tappet lever shaft and remove the tappet lever shaft cap nuts from side of head. Withdraw tappet lever shaft from side of head and tappet levers and push rods from the front. Remove crankcase cover. Remove clutch and belt pulley unit as outlined in paragraph 180. Remove the reduction gear cover and the first reduction gear. Remove cam follower shaft cover (25— Fig. JD37), pull shaft from right side of crankcase and withdraw followers through crankcase cover opening.

Reinstall in reverse order. After cam follower shaft cover and reduction gear are installed, bend the oil deflector fin on side of cam follower shaft cover (if necessary) so that fin covers face of gear. Adjust tappets to 0.015 hot.

ROD AND PISTON UNITS

Series A-B-G-H-Model D

87. Connecting rod and piston units can be removed from front of engine after removing the cylinder head as outlined in paragraph 35, 39 or 41 and the crankcase cover as outlined in paragraph 115, 116 or 117.

NOTE: If cast-in type rod bearings are to be adjusted, do so before removing the connecting rod and piston units. Refer to paragraph 91 for rod bearing adjustment.

Install connecting rods and caps with numbers facing up. Number one cylinder is on left side of tractor. Tighten connecting rod bolts to a torque of 100 ft.-lbs. for series A and D; 62.5 ft.-lbs. for series B, G and H.

PISTONS AND RINGS

Series A-B-G-H-Model D

88. Cast iron pistons are available in standard and 0.045 oversize for series A, B and G; standard, 0.045 and 0.090 oversizes for model D; and standard and 0.030 oversize for series H.

Check pistons, piston rings and cylinders against the values listed below.
Piston Skirt Clearance

Series A	0.005-0.008
Series B (prior 201000)	0.004-0.007
Series B (201000 and up)	0.005-0.009
Series G	0.006-0.009
Series H	0.003-0.006
Model D	0.006-0.010

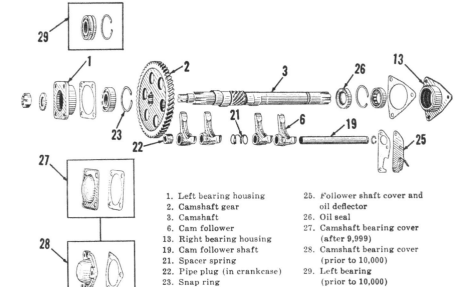

1. Left bearing housing
2. Camshaft gear
3. Camshaft
6. Cam follower
13. Right bearing housing
19. Cam follower shaft
21. Spacer spring
22. Pipe plug (in crankcase)
23. Snap ring
25. Follower shaft cover and oil deflector
26. Oil seal
27. Camshaft bearing cover (after 9,999)
28. Camshaft bearing cover (prior to 10,000)
29. Left bearing (prior to 10,000)

Fig. JD37—Exploded view of series H camshaft and associated parts. Gear backlash is controlled by the position of left bearing housing (1) on studs. Bearing bore in housing (1) is eccentric with respect to housing bore in crankcase.

Piston Skirt Diameter
  Series A 5.493-5.494
  Series B (prior 201000)..4.494-4.495
  Series B (201000 and up) 4.681-4.683
  Series G 6.117-6.118
  Series H 3.557-3.559
  Model D 6.740-6.742
Cylinder Bore
  Series A 5.499-5.501
  Series B (prior 201000)..4.499-4.501
  Series B (201000 and up) 4.688-4.690
  Series G 6.124-6.126
  Series H 3.562-3.563
  Model D 6.748-6.750
Compression Ring End Gap
  Series A-G 0.039-0.053
  Series B 0.025-0.035
  Series H 0.015-0.025
  Model D 0.030-0.040
Oil Ring End Gap
  Series A-G 0.028-0.040
  Series B 0.020-0.030
  Series H 0.015-0.025
  Model D 0.039-0.053
Suggested maximum compression
  ring side clearance...........0.005
Suggested maximum oil ring side
  clearance 0.004

PISTON PINS AND BUSHINGS
Series A-B-G-H-Model D

90. The full floating type piston pins are retained in the piston pin bosses by snap rings, and are available in standard, 0.003 and 0.005 oversizes for series A, B, G and model D; standard and 0.005 oversize for series H.

Piston pin should have a thumb push fit in the unbushed piston and have a clearance of 0.0002-0.002 in the connecting rod bushing. Check the piston pin against the values listed below.

Piston pin diameter
  Series A-G 1.749-1.750
  Series B 1.416-1.417
  Series H 1.104-1.105
  Model D 1.749-1.750

CONNECTING RODS AND BEARINGS
Series A-B-G

91. Early production rod bearings are of the shimmed type and can be adjusted by varying the number of shims between the rod and the cap. Actual adjustment is best accomplished by using Plastigage clearance measuring threads and obtaining approximately 0.003 running clearance. Check the crankshaft crankpin diameter against the values listed below.

Crankshaft crankpin diameter
  Series A 2.998-3.000
  Series B 2.749-2.750
  Series G 3.374-3.375

Series G tractors after Ser. No. 63924 were factory equipped with slip-in, precision type rod bearings. Series A tractors 488000 and up, series B 96000 and up and all series G can be equipped with precision type rod bearing inserts providing the new type connecting rods are also installed.

Bearing inserts are available in undersizes of 0.002-0.004.

Model D

94. On model D tractors prior to 155345, connecting rod bearings are of the bronze-backed, shimmed, babbitt lined type which can be renewed without removing connecting rods from tractor. Adjust connecting rod bearings as outlined in paragraph 91.

On model D tractors after 155344, connecting rod bearings are of the shimmed spun-babbitt type. Renewal of rod bearings is accomplished by removing rods and installing re-babbitted connecting rods. Adjust connecting rod bearings as outlined in paragraph 91.

Crankshaft crankpin diameter is 3.499-3.500.

Series H

96. On series H1000 to H21499 tractors, connecting rod bearings are shimless steel-backed, babbitt lined, non-adjustable, slip-in, precision type which can be renewed without removing connecting rods from tractor. For working clearance, however, it is necessary to remove the tappet lever push rods from tractor. Crankshaft crankpin diameter is 2.061-2.063. Bearing inserts are available in standard and 0.005 undersize.

NOTE: Connecting rods can be replaced with the later type, which have spun babbitt, shimmed bearings.

On series H tractors after 21499, connecting rod bearings are of the spun-babbit shimmed type. Renewal of rod bearings is accomplished by removing rods and installing rebabbitted connecting rods. Adjust connecting rod bearings as outlined in paragraph 91.

CYLINDER BLOCK
Series A-B-G

98. To remove cylinder block, first remove connecting rod and piston units as outlined in paragraph 87 and proceed as follows: Remove upper water pipe and on A series tractors 584000 and up, remove the two studs from right side of upper water pipe flange on cylinder block. On A series tractor prior to 584000 and B series tractors prior to 201000, it is necessary to loosen fan shaft front and rear bearings or in some individual cases, remove governor and fanshaft before upper water pipe can be removed. Also, on A series tractors prior to 584000 and B series tractors prior to 201000, disconnect the tappet lever oiler tube from rear of cylinder block. On A series tractors 584000 and up, B series tractors 201000 and up remove spark plugs and compression release cocks. On G series tractors, it is advisable to remove governor and fan shaft assembly. On B series tractors 201000 and up, loosen spark plug

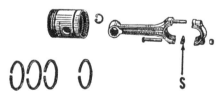

Fig. JD43—Exploded view of series B connecting rod and piston assemblies. Series A & G is similar except five rings are used. Adjustable type connecting rod bearings are adjusted by shims (S).

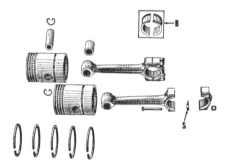

Fig. JD44—Exploded view of model D connecting rod and piston assembly. Model D prior to Ser. No. 155345 was equipped with bronze-backed, insert type bearings (B). Bearings are adjusted by shims (S).

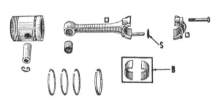

Fig. JD46—Exploded view of series H connecting rod and piston assemblies. Steel-backed, babbitt lined, non-adjustable, precision type bearing inserts (B) were used on tractors prior to 21,500. Tractors after 21,499 were equipped with spun-babbitt type bearings which are adjusted with shims (S). Notice that connecting rod upper bolts are installed with nut toward piston.

wire conduit and move wires out of way. Remove stud nuts retaining cylinder block to crankcase, and on tractors so equipped, remove cap screws retaining cylinder block to frame.

Slide cylinder block forward and off studs, tilt block as required and lift block out over right side of frame. On late model B tractors, the following procedure may be helpful in removing the cylinder block. Turn block so that head studs are down through bottom opening of frame and upper water pipe flange is toward left side of frame. Lift block up until top row of studs is above lower flange of frame left side and tip block against upper flange of frame right side. Then, while holding block at this angle, lift block up and out over frame right side.

Reinstall cylinder block by reversing the removal procedure and, if governor was removed, refer to paragraph 142 for governor installation procedure. On B series tractors 201000 and up, install copper washer on left front cap screw of upper water pipe flange.

Model D

100. To remove cylinder block, remove connecting rod and piston units as outlined in paragraph 87 and remove air cleaner and heat indicator sending unit. Remove hood and fuel tank. Detach fan shaft rear bearing and fan shaft front bearing retainer and slide the fan shaft forward. Remove upper water pipe.

Support front of tractor under crankcase, detach front support from bottom of cylinder block, remove cylinder block retaining stud nuts and slide block from studs. See Fig. JD48.

Series H

101. To remove cylinder block, remove connecting rod and piston units as outlined in paragraph 87, disconnect tappet lever oiler tube from inside of crankcase and remove upper water pipe. Remove spark plugs and shields. Disconnect fuel tank rear bracket from cylinder block, remove stud nuts and cap screws retaining block to crankcase and frame, and slide cylinder block off studs.

CRANKSHAFT AND MAIN BEARINGS

Series A-B-G

On series A (prior to 694828), B and G (prior to 63384), the split type main bearing shells are carried in main bearing housings and shims are provided for adjustment.

On series A (after 694827) and G (after 63383), main bearing bushings are of the aluminum alloy sleeve type which are pressed into the main bearing housings. Main bearing bushings are not available as individual parts, but as a unit with the main bearing housings, on a factory exchange basis. The exchange main bearing and housing units are pre-sized.

Field installation of the sleeve type bearing and housing units can be made on early production A tractors, providing the crankshaft is not excessively worn.

102. **LEFT MAIN BEARING—ADJUST.** To adjust the left main bearing on series A (prior 694828), B and G, remove flywheel as outlined in paragraph 110. Remove left main bearing cover (2—Fig. JD50). Punch mark the flywheel spacer and crankshaft to provide an index mark for correct in-

stallation and remove flywheel spacer (3). Remove pipe plugs (5) from main bearing housing. Remove main bearing adjusting bolts (6) and vary number of shims (9) to obtain not less than 0.002 and not more than 0.006 clearance. An accurate and fast method of determining the running clearance is by the use of Plastigage clearance measuring threads.

When reassembling, do not tighten the bearing adjusting bolts until after shims have been positioned against crankshaft. Renew seal (4) in flywheel spacer (refer to paragraph 109) and reassemble balance of parts by reversing the disassembly procedure making certain that the main bearing housing cover is centered about the crankshaft.

102A. **RIGHT MAIN BEARING—ADJUST.** To adjust the right main bearing, remove the belt pulley and clutch assembly as outlined in paragraph 177. Remove right brake and reduction gear cover. Remove the main bearing adjusting bolts and vary the number of shims (9) to obtain not less than 0.002 and not more than 0.006 clearance which can be accurately measured by using Plastigage thread.

When reassembling, do not tighten the bearing adjusting bolts until inner edge of shims has been positioned against crankshaft journal. These shims should be against the journal to prevent the passage of an excessive quantity of oil into the pulley assembly.

102B. **MAIN BEARINGS—RENEW.** To renew the main bearings on series A (prior 694828), B and G, it is necessary to remove the main bearing housings from tractor. To remove the

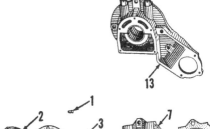

1. Spacer drive pin
2. Left bearing cover
3. Flywheel spacer
4. Cork seal
5. Pipe plug
6. Bearing bolt
7. Left bearing housing
8. Left bearing insert
9. Shims
10. Crankshaft
11. Right bearing housing
12. Right bearing dowel pin
13. Left bearing housing with starter bracket (G 26000 and up only)
14. Right bearing insert

Fig. JD48—When removing cylinder block from model D, support front of tractor under main case and detach front support from cylinder block.

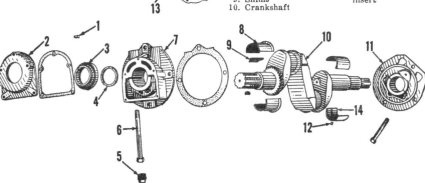

Fig. JD50—Exploded view of series G crankshaft and main bearings. Series A (prior 684828) and B are similar except pipe plugs (5) are at front of housing (7) instead of at rear. Main bearings on series A after 684827 are aluminum full sleeve type.

main bearing housings, follow the adjustment procedure given in the preceding paragraph. In addition remove crankcase cover and disconnect oil lines from housings and remove the housing retaining cap screws.

To renew the main bearing and housing units on A series tractors after 694827, follow the procedure given for series A (prior 684828), B and G, except renew the bearing and housing units, which are available on a factory exchange basis.

Main bearing diametral clearance of sleeve type bearings is 0.004-0.006.

103. CRANKSHAFT. To remove the crankshaft, remove crankcase cover, disconnect oil lines at main bearing housings and remove connecting rod bearing caps. Remove main bearing housings and withdraw crankshaft through side of crankcase. Check the crankshaft against the values listed below.

Crankshaft main journal diameter
Series A 2.748-2.750
Series B 2.248-2.250
Series G 2.998-3.000
Pulley journal diameter
Series A 2.245-2.247
Series B (prior 201000)..1.745-1.747
Series B (after 200999)..1.994-1.996
Series G 2.998-3.000
Crankshaft crankpin diameter
Series A 2.998-3.000
Series B 2.749-2.750
Series G 3.374-3.375

Make certain that crankshaft oil passages are clean and install crankshaft by reversing the removal procedure. Mesh camshaft gear and crankshaft gear so that timing marks are in register.

Model D

105. MAIN BEARINGS—ADJUST. To adjust main bearings, remove crankcase cover as outlined in paragraph 116. Remove the flywheel, left hand main bearing housing cover and flywheel spacer. Remove the clutch, belt pulley and reduction gear cover. Loosen bearing housing inner bolts, remove outer bolts and vary the num-

ber of shims (9—Fig. JD52) to obtain desired 0.002-0.006 clearance which can be easily measured with Plastigage thread.

105A. MAIN BEARINGS—RENEW. To renew the main bearings, remove the crankcase cover and disconnect oil lines from the main bearing housings. Remove the flywheel, left hand main bearing housing cover and flywheel spacer. Remove clutch and belt pulley as outlined in paragraph 177 and the reduction gear cover. Unbolt the main bearing housing from the main case, remove the housings and install the bearing shells.

Reassemble the parts by reversing the disassembly procedure and refer to paragraph 109.

106. CRANKSHAFT. To remove crankshaft, remove crankcase cover as outlined in paragraph 116 and remove connecting rod caps and main bearing housings (refer to paragraph 105A). Work left end of crankshaft through left main bearing housing opening in crankcase. Raise right end of crankshaft up through right rear of crankcase cover opening. Then turn crankshaft into position so that counterbalance will clear and worm the crankshaft out through left main bearing opening. Check crankshaft against the values listed below.

Main journal diameter..2.998-3.000
Pulley journal diameter 2.998-3.000
Rod journal diameter...3.499-3.500

Make certain that crankshaft oil passages are clean and install crankshaft by reversing the removal procedure. Mesh crankshaft gear and camshaft gear so that timing marks are in register.

Series H

107. MAIN BEARINGS. Bearings are one-piece, non-adjustable precision type bushings and are supplied, on the exchange basis, installed in the main bearing housings (7 and 11 — Fig. JD53). To renew the main bearings, remove crankcase cover as outlined in paragraph 117 and remove the main bearing oil lines. Remove flywheel as outlined in paragraph 110. Remove main bearing cover (2) and flywheel spacer (3). Remove the left main bearing housing. Remove the clutch and belt pulley as outlined in paragraph 180 and reduction gear cover. Remove the right main bearing housing.

Reassemble in reverse order, making certain that crankshaft gear and camshaft gear timing marks are in register and refer to paragraph 109.

108. CRANKSHAFT. To remove crankshaft, remove main bearing housings as in paragraph 107. Remove tappet cover (rocker arm cover), loosen tappet adjusting screws and remove push rods. Disconnect connecting rods from crankshaft and withdraw crankshaft from crankcase. Check crankshaft against the values listed below.

Main journal diameter..2.061-2.062

Rod journal diameter...2.061-2.062

Make certain that crankshaft oil passages are clean and install crankshaft by reversing the removal procedure. Mesh crankshaft gear and camshaft gear so that timing marks are in register.

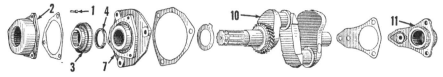

Fig. JD53—Exploded view of series H crankshaft and main bearings. The sleeve type bearings are supplied on the exchange basis installed in the bearing housings. Refer to Fig. JD50 for legend.

MAIN BEARING OIL SEAL

Series A-B-G-H-Model D

109. Left main bearing oil seal (4—Fig. JD50, 52 & 53) is located in flywheel spacer (3), between cover and left main bearing. To renew oil seal remove flywheel as outlined in paragraph 110. If crankshaft is equipped with flywheel locating screw, remove the screw. Remove main bearing cover. Punch mark flywheel spacer and crankshaft to provide an index mark

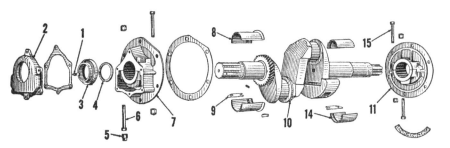

Fig. JD52—Exploded view of model D crankshaft and main bearings. The bearing inner bolts are shown at (15). Refer to Fig. JD50 for legend.

for reinstallation to the same position on crankshaft. Remove flywheel spacer and remove seal from spacer. Carefully insert new seal and reinstall on crankshaft being careful not to damage seal by turning on splines. Align previously affixed punch marks. Reinstall locating screw in crankshaft and reinstall flywheel making certain that driving pin in flywheel is in notch in spacer.

FLYWHEEL AND END PLAY

Series A-B-G-H-Model D

110. To remove flywheel, remove cover on models so equipped, loosen hub bolts and slide flywheel off shaft. When reinstalling flywheel make certain that pin in flywheel aligns with notch in spacer and flywheel is installed in correct position on shaft and securely tightened. Deep keyway of flywheel should be placed over locating screw on shaft; or if wheel hub carries a rivet, the rivet should be engaged with cutaway spline of shaft. On series H and A tractors after Ser. No. A617573, the "V" mark on flywheel should register with "V" mark on shaft. Position of flywheel on crankshaft controls crankshaft end play; move flywheel in or out on shaft to obtain a minimum of end play (0.005-0.010) without causing binding.

Starter ring gear can be driven off flywheel and a new one shrunk on. A mild application of heat on the gear is sufficient to provide the required expansion. Install gear with beveled edge of teeth towards crankcase on A, B, D and G tractors and away from crankcase on the H tractors.

CRANKCASE COVER

Series A-B-G

115. To remove the crankcase cover on A series tractors prior to 584000, B series tractors prior to 201000 and G series tractors, remove starter (if so equipped). To remove starter, remove sides and cover of battery box and disconnect battery ground cable. Remove crankcase breather from cover. Remove cap screws retaining starter to cover, disconnect wire and cable and remove starter. Remove cap screws retaining crankcase cover to crankcase and remove cover.

To remove crankcase cover on series A 584000 and up, and B 201000 and up, remove the cover retaining cap screws and remove the cover.

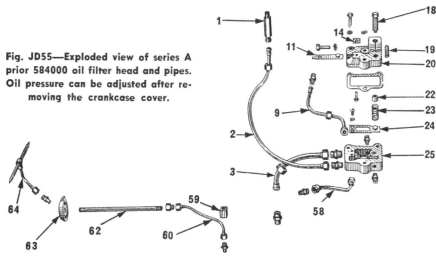

Fig. JD55—Exploded view of series A prior 584000 oil filter head and pipes. Oil pressure can be adjusted after removing the crankcase cover.

1. Governor case connector	19. Spring	25. Filter head
2. Governor oil line	20. Filter head	58. Left main bearing oil pipe
3. Pump discharge pipe	22. Bushing	59. Coupling
9. Right main bearing oil pipe	23. By-pass valve spring	60. Tappet lever oiler pipe
11. Regulating valve spring	24. By-pass leaf spring	62. Oil pipe
14. Clip		63. Oil outlet washer
18. Pressure regulating screw		64. Tappet lever pipe

Model D

116. To remove crankcase cover, disconnect battery ground cable, remove starter retaining cap screws, disconnect cable and remove starter. Remove cap screws retaining cover to crankcase and remove cover.

Series H

117. To remove crankcase cover, disconnect battery ground cable and remove cap screws retaining starter bracket. Disconnect cable from starter and remove starter and bracket. Remove cap screws retaining cover to crankcase and remove cover.

OIL PRESSURE

Series A (Prior 584000)-B (Prior 201000)-G

120. To adjust oil pressure, remove crankcase cover as outlined in paragraph 60 and turn adjusting screw (18—Fig. JD55), (10—Fig. JD56) or (14—Fig. JD57) to obtain the recommended oil pressure of 10-15 psi.

Model D

121. To adjust oil pressure, remove crankcase cover. Remove lock screw and sheet metal shield (27—Fig. JD58) from top of filter head. Turn adjusting screw (14) to obtain the recommended oil pressure of 10-15 psi.

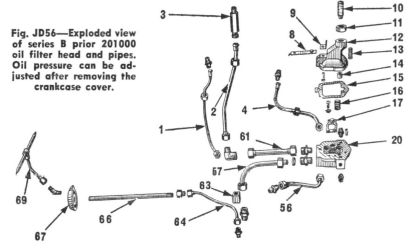

Fig. JD56—Exploded view of series B prior 201000 oil filter head and pipes. Oil pressure can be adjusted after removing the crankcase cover.

1. Camshaft bearing oil line	12. Filter head cover	57. Pump discharge pipe
2. Governor oil line	13. Spring	61. Oil line to camshaft and indicator
4. Right main bearing oil line	14. Bushing	
8. Leaf spring	16. Spring	64. Tappet oil line
9. Clip	17. By-pass valve	66. Oil pipe
10. Oil pressure regulating screw	20. Filter head	67. Outlet washer
11. Jam nut	56. Left main bearing oil line	69. Tappet oil line

1. Pump discharge pipe
2. Left main bearing oil line
3. Governor oil pipe
4. Tappet oiler pipe
5. Connector
8. Filter head cover
9. Bushing
10. Leaf spring
11. Spring
14. Pressure adjusting screw
16. Dowel pin
17. Spring
18. Leaf spring
19. Filter head
20. Right main bearing oil line
24. Jam nut
32. Coupling

Series H-A (after 583999)-B (after 200999)

122. To adjust the oil pressure, remove pipe plug or cap nut from right side of crankcase under pulley. Turn adjusting screw (14—Fig. JD59 or Fig. JD60) clockwise to increase oil pressure. DO NOT ADJUST WITH ENGINE RUNNING. Pressure should be between 10-15 pounds with engine at operating temperature.

OIL FILTER BODY
Series A-B-G-H-Model D

Zero or low oil pressure can be caused by a distorted oil filter body (14—Fig. JD62 or 10—Fig. JD63). Distortion of filter body is usually caused by over tightening of the lower cover nut. This nut should be tightened only enough to eliminate leakage of oil at this point. If leakage occurs with normal tightening, renew rubber gasket, but do not overtighten the nut.

125. **RENEW.** To remove the oil filter body, remove the filter element and crankcase cover. Disconnect oil lines from filter head and remove the filter head and filter outlet assembly.

If difficulty is encountered in removing the oil filter outlet, use a hook tool as shown in Fig. JD64. Slip hooks of tool into holes of filter outlet (9—Fig. JD63 or 11—Fig. JD62) and hammer on inside of loop while holding hooks into holes. On all except H, place wooden block in filter body and jack up block forcing body out of crankcase recess. On series H, use wooden block and drive the body rearward. Remove filter body through crankcase cover opening.

To install oil filter body, coat the portion below the "bead" with white lead or equivalent sealing compound to prevent leaks and facilitate drawing body into recess of crankcase. Place filter head on the body to make certain that cap screw holes in body and head are aligned and that oil lines will align with filter head connections; then, remove filter head. To draw body into crankcase, use a long bolt and two steel plates as shown in Fig. JD65. Tighten the nut until bead of body seats against crankcase. Install remainder of parts by reversing the removal procedure.

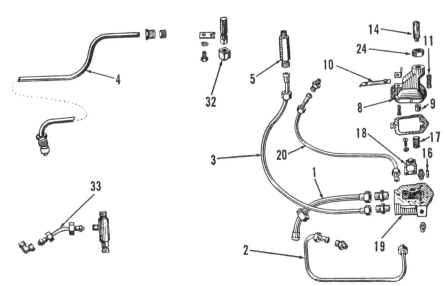

Fig. JD57—Exploded view of series G oil filter head and pipes. Oil pipe (33) was used prior to Ser. No. 13910.

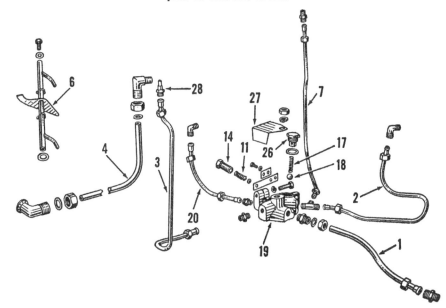

Fig. JD58—Model D oil filter head and pipes. By-pass valve plug (26) is located under shield (27). Oil supply to governor is limited by oil jet (28). Refer to legend for Fig. JD59.

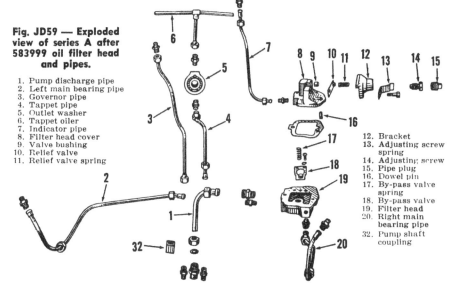

Fig. JD59 — Exploded view of series A after 583999 oil filter head and pipes.

1. Pump discharge pipe
2. Left main bearing pipe
3. Governor pipe
4. Tappet pipe
5. Outlet washer
6. Tappet oiler
7. Indicator pipe
8. Filter head cover
9. Valve bushing
10. Relief valve
11. Relief valve spring
12. Bracket
13. Adjusting screw spring
14. Adjusting screw
15. Pipe plug
16. Dowel pin
17. By-pass valve spring
18. By-pass valve
19. Filter head
20. Right main bearing pipe
32. Pump shaft coupling

OIL PUMP

Series A-B-G-H

127. **R&R AND OVERHAUL.** To remove the oil pump, remove crankcase cover and disconnect oil lines from pump body. On H series tractors, remove pipe plug from right side of crankcase and screw the oil pressure adjusting screw completely in. Remove cap screws (or nuts) retaining pump to crankcase and withdraw pump from crankcase. Install pump be reversing the removal procedure.

128. Disassembly of the pump is evident after an examination of the unit and reference to Figs. JD 63 or 66. Gasket between bottom cover and pump body prevents leakage at this point and provides end clearance for the gears. On series A, B and G, renew the oil pump drive coupling (13—Fig. JD63).

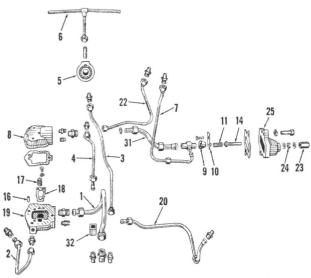

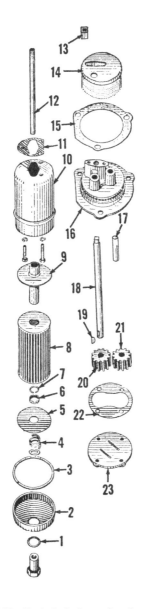

Fig. JD60—Exploded view of series B after 200,999 oil filter head and pipes.

1. Pump main discharge pipe
2. Left main bearing pipe
3. Governor pipe
4. Tappet pipe
5. Outlet washer (connector)
6. Tappet oiler
7. Indicator pipe
8. Filter head cover
9. Valve bushing
10. Relief valve
11. Relief valve spring
14. Adjusting screw
16. Dowel pin
17. By-pass valve spring
18. By-pass valve
19. Filter head
20. Right main bearing pipe
22. Camshaft bearing pipe
23. Cap nut
24. Lock nut
25. Relief valve housing
31. Relief pipe
32. Pump shaft coupling

Fig. JD61—Series B (201-000 and up) oil pressure adjustment. Turn the screw "in" to increase the oil pressure.

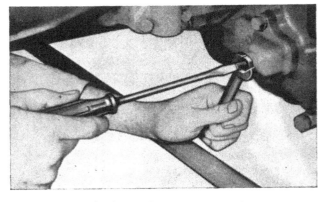

2. Gasket
3. Cover
5. Gasket
6. Spring
7. End plate
8. Spacer
9. Snap ring
10. Filter element
11. Filter outlet
14. Filter body
15. Gasket
16. Stud
17. Filter head
18. By-pass leaf spring
21. Gasket
23. Cover
24. Spring

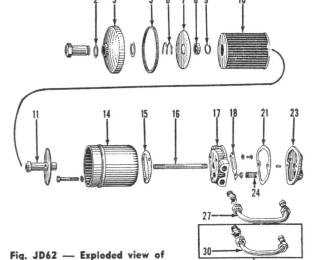

Fig. JD62 — Exploded view of series H oil filter and filter head. The various discharge pipes which have been used are shown at (27, 30 & 35).

Fig. JD63—Exploded view of series B oil pump and filter assembly. Series A and G pumps and filters are similar. Series D filter is similar.

1. Gasket	12. Filter stud
2. Filter bottom	13. Shaft coupling
3. Bottom gasket	14. Oil strainer
4. Spring	15. Pump body
5. End plate	16. Pump body
6. Spacer	17. Idler gear shaft
7. Snap ring	18. Pump shaft
8. Filter element	19. Shaft key
9. Filter outlet	20. Pump gear
10. Filter body	21. Idler gear
11. Top gasket	22. Cover gasket
	23. Pump cover

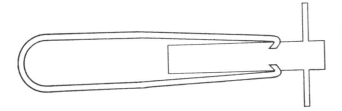

CARBURETOR

Series A-B-G-H-Model D

133. Marvel-Schebler carburetors, series DLTX are used. Refer to carburetor for model number and to Appendix 1 on Page 78 of this manual for calibration and adjustment data.

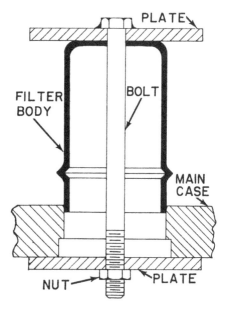

Fig. JD65—Suggested home-made tool for installing Deere oil filter body. The long bolt can be welded to the upper plate.

Model D

129. To remove oil pump, remove flywheel as outlined in paragraph 110. Remove strainer assembly (14—Fig. JD67) from left side of crankcase. Remove cap screws retaining pump to side of crankcase and remove pump.

Disassembly of the pump is evident after an examination of the unit and reference to Fig. JD67. Gasket (22) between cover and pump body prevents leaks at this point and provides end clearance for gears.

Install pump by reversing the removal procedure.

MANIFOLDS

Series A-B-G-H-Model D

132. To remove manifolds, remove cylinder head as outlined in paragraph 35, 39 or 41. Carbon can be removed from intake manifold by directing flame of gasoline blow-torch into manifold and "cooking" carbon loose. If manifold retaining studs are removed on series B, make certain that they are reinstalled properly. If wrong end of stud is screwed into cylinder head, it may damage tappet rod (push rod) sleeves.

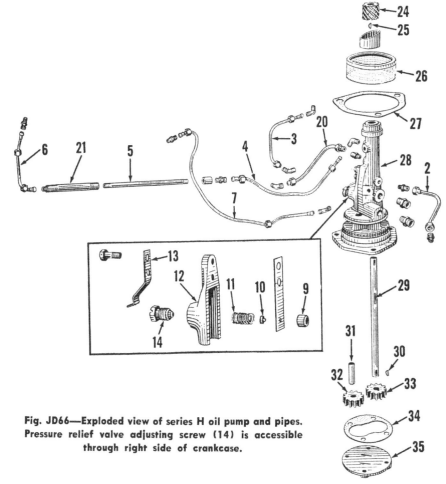

Fig. JD66—Exploded view of series H oil pump and pipes. Pressure relief valve adjusting screw (14) is accessible through right side of crankcase.

2. Left main bearing pipe	10. Relief valve	20. Right main bearing pipe
3. Governor pipe	11. Relief valve spring	21. Lead stud
4. Tappet pipe (rear)	12. Bracket	24. Helical drive gear
5. Lead pipe	13. Adjusting screw spring	25. Gear key
6. Tappet pipe (front)	14. Adjusting screw	27. Pump gasket
7. Gauge pipe		28. Pump body
9. Valve bushing		29. Pump shaft
		30. Shaft key
		31. Idler gear shaft
		32. Idler gear
		33. Pump gear
		34. Cover gasket

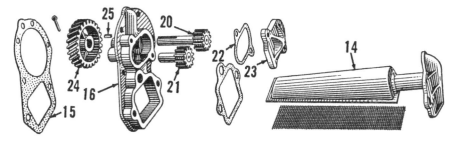

Fig. JD67—Exploded view of model D oil pump assembly.

14. Oil strainer	20. Pump gear	23. Pump cover
15. Pump gasket	21. Idler gear	24. Helical drive gear
16. Pump body	22. Cover gasket	25. Gear key

GOVERNOR

ADJUSTMENT

Series A-B-G-H-Model D

140. SPEED ADJUSTMENT. On series A, B, G and H, turn adjusting screw at steering column to limit engine speed to rpm specified below. Speed adjustment on model D is by changing the position of the sleeve (S—Fig. JD69A) on the throttle rod.

Engine high idle (no-load rpm).

Series A	1080
Series B (prior 201000)	1340
Series B (after 200999)	1370
Series G	1115
Series H (crankshaft)	1540
Series H (belt pulley)	770
Model D	980

NOTE: On series A, B and G and model D, the belt pulley speed is the same as the engine speed.

Engine load rpm

Series A	975
Series B (prior 201000)	1150
Series B (after 200999)	1250
Series G	975
Series H (crankshaft)	1400
Series H (belt pulley)	700
Model D	900

See note above.

141. LINKAGE ADJUSTMENT. Free-up and align all linkage to remove any binding. Adjust or renew any parts causing lost motion. Disconnect throttle rod at carburetor. With hand control in full speed position, adjust length of carburetor throttle rod so that rod is ½ hole short from entering hole in carburetor throttle shaft lever when carburetor butterfly is in wide open position.

REMOVE AND REINSTALL

Series A-B-G

142. On series A prior 584000, B prior 201000 and G, remove starter. Remove generator if tractor is so equipped. Remove exhaust pipe and governor controls. Remove screws from front of hood. Remove bolts attaching fuel tank rear bracket to governor case. Remove oil pipe to tappet cover and on A series tractors prior to 584000, B series tractors prior to 201-000 and all G series tractors, remove crankcase cover and disconnect oil line to governor case. Remove magneto or distributor. Remove cap screws retaining governor case to crankcase, noting location of dowel-type cap screw (1—Fig. JD70) which is used on some models. Remove fan front bearing housing cap screws. On models so equipped, remove ventilator pipe from governor housing. Support fuel tank and hood assembly on block or jack, remove fuel filter bowl and lift governor and fan assembly from tractor.

NOTE: On A series tractors 584000 and up and B series tractors 201000 and up, it is usually necessary to remove the steering shaft before the hood and fuel tank can be raised; also, it is necessary to remove ventilator pipe from fan front bearing housing before removing governor assembly.

When reinstalling governor, make certain that timing marks on the governor drive gear and camshaft gear are in register. If marks are obliterated, make certain that flywheel is correctly installed as outlined in paragraph 110, then turn flywheel until mark "L.H. IMPULSE" on flywheel rim aligns with mark on gear cover or gear case. Turn governor shaft until groove in magneto or distributor drive coupling flange is horizontal and mesh gears. Balance of reinstallation is reverse of removal. Time the magneto or distributor as outlined in paragraph 162 or 163.

Model D

145. Remove radiator as outlined in paragraph 151; then, remove fuel tank. Remove magneto. Remove governor controls and ventilator tube from governor to air cleaner. Remove fan assembly and oil line to tappet oiler. Detach fan rear bearing housing and remove cap screws from fan front bearing housing. Slide fan shaft forward, remove governor case retaining cap screws and remove governor.

Reinstall governor by reversing the removal procedure and mesh governor gear with idler gear when flywheel mark "L.H. IMPULSE" aligns with mark on gear cover or gear case and the drive slots in the magneto drive coupling are horizontal. Retime magneto as outlined in paragraph 162.

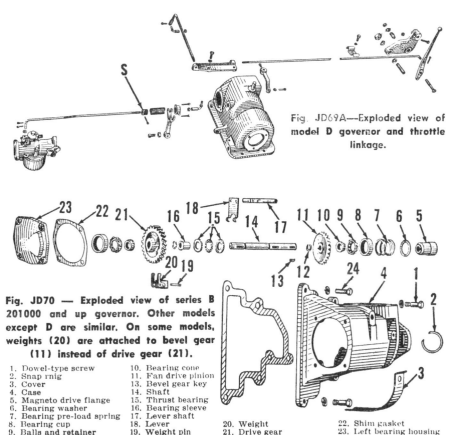

Fig. JD69A—Exploded view of model D governor and throttle linkage.

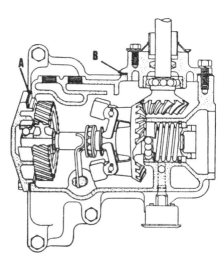

Fig. JD69—Sectional view of typical governor assembly. Bevel gear mesh and backlash is controlled by shim gaskets (A) and (B).

Fig. JD70 — Exploded view of series B 201000 and up governor. Other models except D are similar. On some models, weights (20) are attached to bevel gear (11) instead of drive gear (21).

1. Dowel-type screw		10. Bearing cone	
2. Snap rnig		11. Fan drive pinion	
3. Cover		13. Bevel gear key	
4. Case		14. Shaft	
5. Magneto drive flange		15. Thrust bearing	
6. Bearing washer		16. Bearing sleeve	
7. Bearing pre-load spring		17. Lever shaft	
8. Bearing cup		18. Lever	
9. Balls and retainer		19. Weight pin	
20. Weight		22. Shim gasket	
21. Drive gear		23. Left bearing housing	

146. IDLER GEAR—R&R. Remove crankcase cover. If additional working clearance is required, remove governor as outlined in paragraph 145. Remove nut from idler gear spindle, then drive spindle into crankcase and withdraw gear (33—Fig. JD72).

Bushing (32) in gear is pre-sized and, if carefully installed, requires no final sizing. Install bushing with chamfered edge towards the right side. When reassembling, make certain that thrust washer (34) is installed between the gear and crankcase and the locating pin (31) is engaged in the notch of the washer and the notch in the crankcase.

Series H

148. Remove starter and generator if tractor is so equipped. Remove exhaust pipe and breather pipe. Remove fuel filter bowl. Remove magneto. Remove cap screws retaining governor case to crankcase and remove governor and fan assembly.

When reinstalling governor, make certain that timing marks on the governor drive gear and camshaft gear are in register. If marks are obliterated, make certain that flywheel is correctly installed as outlined in paragraph 110; then, turn flywheel until mark "L.H. IMPULSE" on flywheel rim aligns with mark on gear cover or gear case. Turn governor shaft until groove in magneto coupling flange is horizontal and mesh gears. Balance of reinstallation is reverse of removal. Time magneto as outlined in paragraph 162.

OVERHAUL
Series A-B-G-H—Model D

A general procedure for overhauling the governor is given in the following paragraphs. Differences in construction are evident after an examination of the unit and reference to Figs. JD70 & 72.

149. Remove left bearing housing (23). Remove set screw from governor arm and remove arm and spring. Governor shaft, gear and bearings assembly can now be removed from case. Use O.T.C. puller or equivalent and remove magneto coupling flange (5) from shaft. Press shaft out of bearing inner races and gears. Renew worn parts and reassemble in reverse order. For good governor action, no play should exist in weights, weight pins, bearings or controls. Be sure drilled oil hole in governor housing is open.

Fan drive gears are supplied in matched pairs only.

When reassembling, vary number or thickness of shims between rear fan bearing housing and governor case so that rear surface of fan shaft gear is flush with edge of bevel gear (11) on governor shaft. Vary number or thickness of shims (22) between left bearing housing and governor case to provide gear backlash of 0.004-0.006. Renew felt seal in fan shaft front bearing and pack bearing with a high melting point lubricant. For fan shaft disassembly, refer to paragraph 155 or 157.

149A. NOTE: On model D, the magneto coupling flange can be removed without removing the governor shaft. Screw cap screws through the flange and against the bearing housing, thereby forcing the flange off of the shaft.

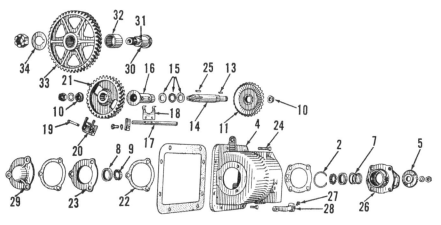

Fig. JD72—Exploded view of model D governor.

2. Snap ring	15. Thrust bearing	25. Drive gear key
4. Case	16. Bearing sleeve	26. Right bearing housing
5. Magneto drive flange	18. Lever	27. Set screw
7. Bearing pre-load spring	19. Weight pin	28. Control lever
8. Cup	20. Weight	29. Bearing cover
9. Ball retainer	21. Drive gear	30. Idler gear spindle
10. Cone	22. Shim gasket	31. Dowel pin
11. Fan drive pinion	23. Left bearing housing	32. Gear bushing
13. Key		33. Idler gear
14. Shaft		34. Thrust washer

COOLING SYSTEM

RADIATOR
Series A-B-G-H

150. REMOVE AND REINSTALL. To remove the radiator, first remove the grille and the steering (worm) shaft. Remove hood and shutter. Remove air cleaner and fan shroud. Disconnect radiator hoses, remove cap screws retaining radiator to frame and remove radiator.

Model D

151. To remove radiator, remove hood and grille assembly. Detach radiator top tank from upper water pipe and lower tank from cylinder head.

To prevent misalignment of radiator flanges and consequent leakage,

do not complete tightening of lower tank to cylinder head until after top tank flange has been tightened.

FAN AND SHAFT
Series A-B-G-H

155. R&R AND OVERHAUL. To remove fan and shaft assembly, remove governor and fan assembly as a unit as outlined in paragraph 142 or 148. Remove cap screws retaining rear fan bearing housing (31—Fig. JD78) to governor case and remove fan shaft assembly from governor case. With shaft in vertical position and using a bar or press, force fan and drive discs into front bearing housing and remove split cone locks (1) and keeper (2)

from front of shaft. Withdraw drive discs, friction washers and fan and front bearing parts from front of tube. Remove shaft, gear and rear assembly from rear of tube. Repack front bearing with high melting point lubricant and renew oil seal when reassembling.

156. CRANKCASE VENTILATING PUMP. Series A (584000-692120) and series B (201000-300866) are equipped with a vane type crankcase ventilating pump which is mounted on the rear end of the fanshaft tube. When overhauling the fanshaft, disassemble the pump and check the component parts against the values as listed.

Eccentric O. D. 1.344-1.348
Roller O. D. 2.308-2.312
Roller I. D. 1.350-1.354
Vane width 0.995-1.000
Vane thickness 0.495-0.500
Recess width in pump body.......
.................0.502-0.504
Bore in pump body.....2.578-2.580

Late production Series A and B tractors are equipped with a rotor type crankcase ventilating pump which is mounted on the rear end of the fan shaft tube. When overhauling the fan shaft, disassemble the pump and check the component parts against the values listed below.

Roller diameter 0.4982-0.4988
Roller length 0.965-0.970
Rotor thickness 0.964-0.965
Pump body thickness..0.9715-0.9735
Rotor diameter 2.425-2.430
Rotor groove diameter...0.500-0.505

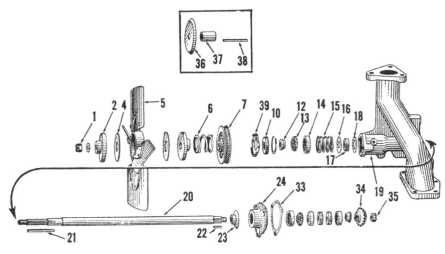

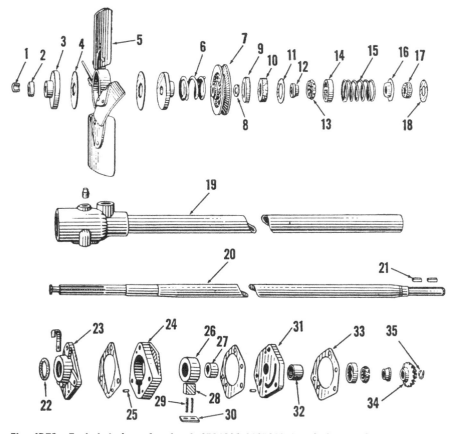

Fig. JD80—Exploded view of model D fanshaft assembly.

2. Friction disc	13. Balls and retainer	20. Fan shaft	33. Shim gasket
4. Friction washer	14. Bearing cup	21. Front key (161,400	34. Bevel gear
5. Fan blades	15. Pre-load spring	up)	36. Dust shield (to
6. Friction spring	16. Washer	22. Gear key	161,399)
7. Pulley	17. Felt seal	23. Dust shield	37. Spacer (to 161,399)
10. Felt seal	18. Washer	24. Rear bearing	38. Front key (to
12. Bearing cone	19. Front brg. housing	housing	161,399)

Model D

157. **R&R AND OVERHAUL.** To remove fan, remove radiator as in paragraph 151. Remove nut (1—Fig. JD80) from front of fan shaft and remove fan. Friction discs and friction washers can then be withdrawn. Front bearing assembly and dust seal can be removed from upper water pipe after removing front bearing cover (39). When reassembling, renew dust seal.

To remove fan shaft assembly, remove governor as in paragraph 145 and withdraw shaft assembly from upper water pipe. When reassembling, tighten fan shaft nut (1) to 10 ft.-lbs. torque.

PRESSURE CAP

Series G

159. The radiator is fitted with a 6-7 lb. pressure cap which raises the coolant boiling point to approximately 230° F.

WATER PUMP

Series A (700200 and up)- B (306600 and up)- G (60700 and up)

160. **R&R AND OVERHAUL.** The V-belt driven, vane type water pump is mounted on the left side of the radiator lower tank as shown in Fig. JD81. Procedure for removing the pump is evident after an examination of the unit.

NOTE: If water is present in the drain hole or slot in bottom of the pump casting, the bellows seal and/or carbon sealing washer should be renewed.

Fig. JD78—Exploded view of series A (584000-692120) fan shaft assembly and ventilating pump. Series B (201000-300866) is similar. Earlier production A and B series tractors and series G and H are similar except the ventilating pump is not used. On late production series A and B tractors, the fanshaft assembly is similar except a rotor type ventilating pump is used.

1. Split cone lock (2)	11. Washer	19. Front bearing	27. Pump eccentric
2. Keeper	12. Bearing cone	housing and tube	28. Pump vane
3. Friction disc	13. Balls and retainer	20. Fan shaft	29. Vane springs
4. Friction washer	14. Bearing cup	21. Key	30. Spring retainer
5. Fan blades	15. Bearing pre-load	22. Packing ring	31. Rear bearing
6. Friction spring	spring	23. Pump cover	housing
7. Generator drive	16. Packing retainer	24. Pump body	32. Spacer
pulley	17. Packing	25. Dowel pin	33. Shim gasket
8. Rubber packing	18. Washer	26. Pump roller	34. Bevel gear
9. Felt retainer			35. Snap ring

To disassemble the pump, refer to Fig. JD82 and proceed as follows: Using a suitable puller, remove the drive pulley from the shaft. Remove the snap ring and press the shaft and bearing assembly out of the impeller and pump housing. The shaft and bearing assembly is available as an assembled unit only.

NOTE: Early production water pumps had a small drain hole drilled into the pump casting; late production pumps have a slot cut into the casting. When overhauling the early production pumps, enlarge the drain hole to ⅜ inch diameter. Also, on early production pumps, the nearest face of the flange on the slinger (shown in Fig.

JD82) was positioned 1 21/32 inches from the end of the shaft. When overhauling the pump, the nearest face of the flange on the slinger should be positioned 1 39/64 inches from the end of the shaft.

To assemble the pump unit, support the pump housing, press the shaft and bearing unit into the housing and install the snap ring. Install the right hand pump attaching cap screw (cap screw which is nearest the water outlet pipe), support shaft from under side and press the drive pulley on the shaft until rear face of pulley is flush with rear end of shaft. Install bellows seal and carbon thrust washer into impeller, making certain that lugs on

the washer are in register with slots in the impeller. Place impeller on a flat surface and press the shaft and housing assembly onto the impeller. Continue pressing the assembly together until highest vane on the impeller is flush with mounting surface of the pump housing. Surface of pump body which contacts the carbon washer must be flat and smooth. If the surface is scored or pitted install a new body or reface same.

NOTE: If any vane protrudes beyond the mounting face of pump housing, the protruding vane will strike the radiator lower tank when pump is installed.

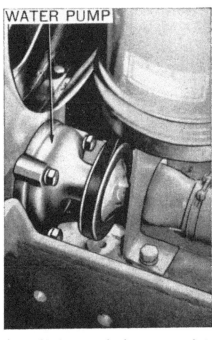

Fig. JD81—Late production A, B and G series tractors are equipped with a coolant circulating pump which is mounted on the radiator lower tank as shown.

Fig. JD82 — Exploded view of A, B and G coolant circulating pump. The shaft and bearings are available as an assembled unit only.

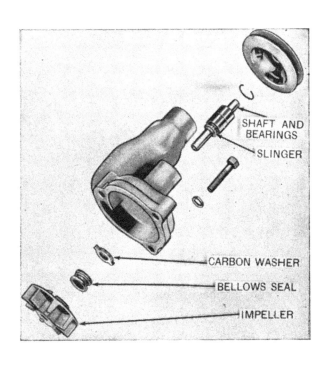

SHAFT AND BEARINGS

SLINGER

CARBON WASHER

BELLOWS SEAL

IMPELLER

IGNITION AND ELECTRICAL SYSTEM

161. Series H, model D and early production series A, B and G tractors are equipped with either Edison-Splitdorf, Fairbanks-Morse or Wico magnetos. Late production series A and G tractors are equipped with a Delco-Remy battery ignition distributor and late production series B with a Wico battery ignition distributor.

Refer to starting motor, generator and regulator for model number and to Appendix 1 of this manual for testing and overhaul data.

MAGNETO TIMING

Series A-B-G-H-Model D

162. Check flywheel for correct position on shaft as outlined in paragraph 110. Turn flywheel until number one (left) cylinder is on compression stroke; then turn slowly until mark "L.H. IMPULSE" on flywheel rim aligns with mark on cover as in Fig. JD85, or mark or dimple on flywheel hub aligns with index mark on cover as in Fig. JD86. Groove in magneto driving flange on governor shaft should be in horizontal position at this time. If groove is not in horizontal position, governor gear (except on D) will have to be retimed to the cam-

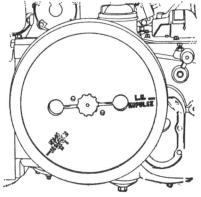

Fig. JD85—Series B ignition timing mark (L.H. IMPULSE) is on rim of flywheel. Series A, D, G and H are similar.

shaft as outlined in paragraph 142 and 148. On D model, coupling flange can be pulled from governor shaft as described in paragraph 149A and re-installed in the correct position.

Turn the magneto shaft to the position where it is firing the front (one nearest cylinder block) terminal. With the magneto shaft in this position and the coupling flange on the governor shaft in the correct position as previously described, install the magneto on the engine and push top of magneto toward front of tractor. Turn fly-wheel backwards about ¼ turn then slowly turn in direction of rotation until flywheel marks again align. Slowly rotate top of magneto rearward until impulse trips and tighten mounting flange.

BATTERY IGNITION TIMING

Series A-B-G

163. Check flywheel for proper installation on the crankshaft as outlined in paragraph 110. Turn flywheel until No. 1 (left) cylinder is coming up on compression stroke and fly-

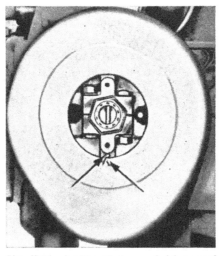

Fig. JD86—Ignition timing mark (depressed dot) on flywheel hub of late production, starter equipped engines.

wheel mark "L.H. IMPULSE" on early production A, B and G tractors, "punch mark on flywheel hub" on late production A, B and G tractors aligns with mark on flywheel cover or gear case. Refer to Figs. JD 85 and 86. At this time, the groove on the distributor

drive flange should be in a horizontal position. If the groove is not in horizontal position, governor gear will have to be retimed to the camshaft as outlined in paragraph 142. Turn the distributor shaft to the position where it is firing the **top** terminal. On some models this position is obtained when the punch marked drive lug is facing front of tractor. On other models, the drive lug is not marked; in which case, on Delco-Remy distributors, remove distributor cap and install the distributor so that rotor arm is facing up. On Wico distributors, remove the distributor cap and install distributor so that rotor arms are facing the 10 and 1 o'clock positions. Install the distributor and tighten the mounting bolts finger tight.

Turn the distributor clockwise as far as it will go, then counter-clockwise until a spark occurs from the **top** terminal. Tighten the distributor mounting bolts and install the spark plug cables.

Running timing is 25-26 degrees B.T.C.

CLUTCH AND BELT PULLEY

ADJUSTMENT

Series A-B-G-H-Model D

170. To adjust the clutch, first remove pulley cover. Remove cotter pins from the three operating bolts and place the clutch control lever in the engaged position. Tighten the nut on each operating bolt about ¼ turn at a time and to the same tension. Refer to Fig. JD90. Check tightness of clutch after each adjustment by disengaging and re-engaging the clutch. When the adjustment is correct, a distinct snap will occur when the clutch is engaged and approximately 40-80 pounds pressure on the series A, B and H and 70-100 pounds pressure on the D and G will be required at the end of the control lever to lock the clutch in the engaged position when the engine is idling.

NOTE: On early production G series tractors which have a cam operated clutch, no snap will occur when clutch goes into engagement. The adjustment, however, will be correct when 70-100 pounds pressure is required at the end of the control lever to lock the clutch in the engaged position when engine is idling.

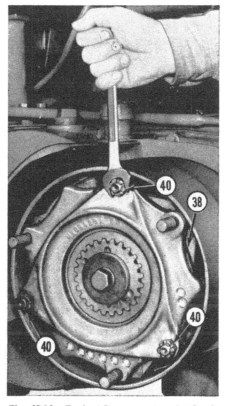

Fig. JD90—Typical Deere tractor clutch adjustment procedure. Tighten each adjusting nut (40) a little at a time until a distinct snap is heard when clutch is engaged. The clutch adjusting disc is shown at (38).

171. To adjust the pulley brake, loosen the adjusting screw lock nut and turn the adjusting screw until the pulley brake will stop the belt pulley when clutch is disengaged and the clutch operating lever is held back. Refer to Fig. JD91.

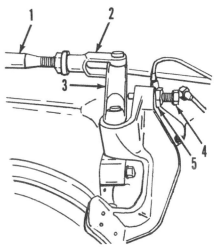

Fig. JD91—Typical Deere tractor clutch linkage and pulley brake adjustment. Differences in construction are evident after an examination of the unit.

1. Rod	4. Adjusting screw
2. Yoke	5. Lock nut
3.	

R&R AND OVERHAUL CLUTCH DISCS

Series A-B-G-H-Model D

173. To remove the clutch discs and facings, remove the pulley cover. Re-move the clutch adjusting disc by re-moving the three adjusting nuts, then remove the clutch release springs. Lift out the linings, discs and/or facings. Remove the clutch drive disc retaining cap screw and on model D, loosen the drive disc clamp screw. Using a suit-able puller if necessary, remove the drive disc. Refer to Figs. JD93 or 94.

174. Observe the following points when reinstalling the discs and fric-tion rings to the pulley and shaft: The tapered or rounded outer edge of the innermost friction ring (free facing) should face inward. The drive disc which is the thickest unlined disc is correctly installed when the "V" marks are in register as shown in Fig. JD94A or when the rivet in the disc hub engages the cutaway spline on the shaft. The rounded outer edge of the outermost friction ring should face outward. On model D, make certain that the drive disc clamp screws are tight. Adjust clutch as outlined in paragraph 170.

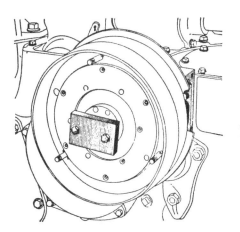

Fig. JD93—Using two half-inch bolts to re-move series A clutch drive disc. The same procedure can be used on series B and G.

Fig. JD94—Using a plate and two cap screws to remove series H clutch drive disc after loosening the disc retaining cap screws.

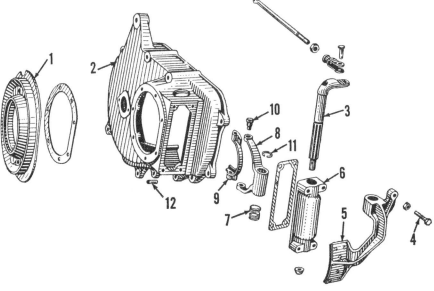

Fig. JD95—Exploded view of series A clutch operating linkage, pulley brake and re-duction gear cover. Series B, D and G are similarly constructed; the differences of which are evident after an examination of the unit.

1. Dust shield	5. Pulley brake	9. Collar
2. Gear cover	6. Fork bearing	10. Pivot (2)
3. Fork shaft	7. Fork spring	11. Snap ring
4. Brake adjusting screw	8. Fork	12. Dowel pin

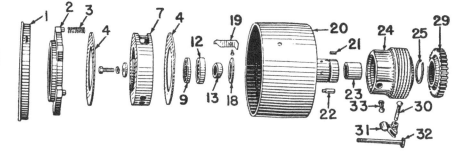

Fig. JD96—Exploded view of series A prior 648000 and B prior 201000 belt pulley and clutch assembly.

1. Cover	18. Bearing washer	24. Operating sleeve
2. Adjusting disc	19. Cover spring	25. Sleeve snap ring
3. Release spring	20. Pulley	29. Pulley gear
4. Free facing	21. Gear key	30. Pin (3)
9. Bearing retainer	22. Sleeve drive pin	31. Dog (3)
12. Pulley bearing	23. Pulley bushing	32. Operating bolt (3)
13. Inner race		33. Toggle (3)

Fig. JD94A—On some Deere tractors, the clutch drive disc is stamped with the letter "V". When installing the disc, the "V" mark must be in register with a similar mark on the crankshaft.

OVERHAUL PULLEY AND SLEEVE
Series A-B-G-Model D

177. To remove the belt pulley assembly, first remove the clutch friction facings and discs as outlined in paragraph 173. Disconnect linkage from the clutch fork shaft, remove the clutch fork bearing retaining cap screws and remove the clutch fork (8—Fig. JD95) and bearing (6) as an assembly. Pulley assembly can then be removed from tractor.

178. Bearing (12—Fig. JD96, 97 or 98) and bushing (23) can be renewed at this time. When renewing bushing (23), carefully press the bushing into pulley and hone or ream the bushing, if necessary, to obtain a free running fit on the crankshaft. When reassembling, make certain that oil holes in pulley hub and operating sleeve are open. Pack the roller bearing with a high melting point grease and reinstall bearing and retainer cup. If bearing is fitted with a **felt** oil seal, as shown in Fig. JD98 do **not** pack with grease.

Belt pulley gear (29) can be removed from pulley by using a suitable puller. The need and procedure for further disassembly is evident after an examination of the unit. To obtain efficient clutch action it is sometimes necessary to renew toggles, pins and clutch dogs. If renewing these parts does not restore original action to an excessively worn unit, machine approximately 0.030 off the outer end of the sleeve on all tractors except series G prior to 13804.

This machining will permit the sleeve to move farther to the right in the clutch engaged position and provide a more satisfactory lock. On the excepted G tractors, which were factory equipped with cam operated type unit, renew the operating sleeve.

NOTE: The early production G tractor cam operated units can be replaced, using the late production, toggle action type.

When reinstalling gear (29) on pulley, heat gear in boiling water and support pulley on hub surface while pressing the gear on.

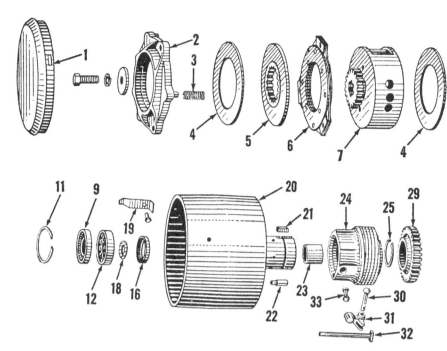

Fig. JD97—Exploded view of series B 201000 and up belt pulley and clutch assembly. Series G and A after 647999 are similar. On series G, bearing (12) is retained by snap ring (11).

2. Adjusting disc	9. Bearing retainer	20. Pulley	25. Sleeve snap ring
3. Release spring	11. Bearing snap ring	21. Gear key	29. Pulley gear
4. Free facing	12. Pulley bearing	22. Sleeve drive pin	30. Pin (3)
5. Sliding drive disc	16. Oil seal	23. Pulley bushing	31. Dog (3)
6. Lined disc	18. Bearing washer	24. Operating sleeve	32. Operating bolt (3)
7. Drive disc	19. Cover spring		33. Toggle (3)

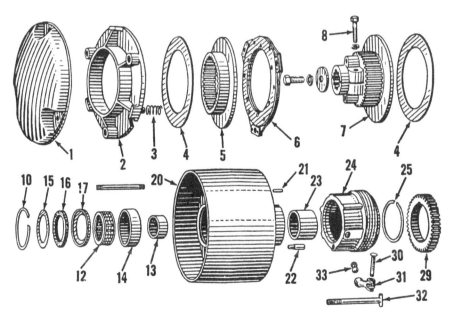

Fig. JD98—Exploded view of model D belt pulley and clutch.

1. Cover	8. Cap screw	16. Oil seal	24. Operating sleeve
2. Adjusting disc	10. Snap ring	17. Oil seal retainer	25. Sleeve snap ring
3. Release spring	12. Pulley bearing	20. Pulley	29. Pulley gear
4. Free facing	13. Inner race	21. Gear key	30. Pin (3)
5. Sliding drive disc	14. Outer race	22. Sleeve drive pin	31. Dog (3)
6. Lined disc	15. Oil seal washer	23. Pulley bushing	32. Operating bolt (3)
7. Drive disc			33. Toggle (3)

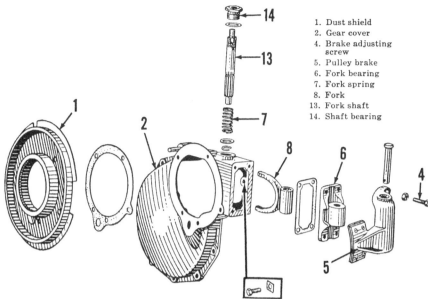

1. Dust shield
2. Gear cover
4. Brake adjusting screw
5. Pulley brake
6. Fork bearing
7. Fork spring
8. Fork
13. Fork shaft
14. Shaft bearing

Fig. JD99—Exploded view of series H clutch operating linkage, pulley brake and reduction gear cover.

from front. Remove clutch discs and facings as outlined in paragraph 173 and remove the pulley. On H tractors prior to 10000, snap ring (10—Fig. JD 100) must be removed before removing the pulley.

181. To overhaul the pulley assembly, follow the same procedure as outlined in paragraph 178. Tractors prior to H 10000 were equipped with a sealed, pre-lubricated bearing (34—Fig. JD100); later tractors were equipped with an open type bearing which must be packed with high melting point grease.

Series H

180. To remove the belt pulley assembly, punch correlation marks on the clutch fork shaft (13—Fig. JD99) and shaft lever for reassembly reference. Remove shaft lever and bearing (6). Remove shaft bearing (14) from top of gear cover, withdraw the fork shaft from above and the fork (8)

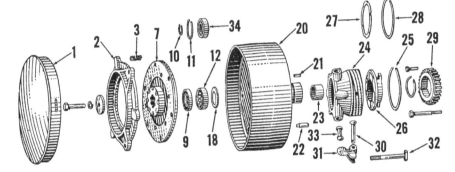

Fig. JD100—Exploded view of series H belt pulley and clutch assembly. Ball bearing (34) and snap rings (10) and (11) were used on tractors prior to 10,000. See legend under Fig. JD97.

TRANSMISSION AND CONNECTIONS

R&R TRANSMISSION COVER

Series A (Prior 584000)- B (Prior 201000)

190. To remove the transmission cover, remove battery sides, cover and battery. Remove choke rod, fuel control rod, shutter control rod and speed control rod. Disconnect heat indicator bulb, oil indicator line and wires to switch and ammeter. Remove bolt from steering shaft upper bearing. Detach bottom of battery box. Remove cap screws retaining cover to transmission case and remove cover.

For procedure for overhaul of the shifter shafts and shifters refer to paragraph 196.

Series A (584000 and up)- B (201000 and up)-G

191. To remove the transmission cover, remove speed control rod and shutter control rod from A and B series tractors; and on G series, remove bolt from steering shaft rear

bearing. Remove cap screws retaining cover to transmission case and remove cover.

For procedure covering overhaul of the shifter shafts and shifters refer to paragraph 196 or 197.

Model D

192. To remove model D transmission covers, first remove hood and dash. Remove cap screws retaining covers to main case and remove covers.

Series H

193. To remove transmission cover, remove steering wheel and speed control rod. Remove cap screws retaining cover to main case and remove cover.

R&R SHIFTER SHAFTS AND SHIFTERS

Series A-B (prior 201000)

196. To remove the transmission shifter shafts and shifters, first remove the transmission top cover as outlined in paragraph 190 or 191 and flywheel as outlined in paragraph 110:

Remove the transmission shaft cover from left side of main case. Move each shifter along the shifter shafts until the shifter pawls rise. Install a cotter pin or nail in the hole in each shifter pawl to lock the pawls in the raised position. Remove the detent plunger (16—Fig. JD116) or (15—Fig. JD117) from the vertical drilled hole which is located to the left of transmission top cover opening. Remove cap screws (1) and adjusting screws (2) from left end of the shifter shafts and on A series prior to 648000 and B series loosen the vertical lock screw (3—Fig. JD116) and remove cotter pin (5). Remove pointed set screw (12). Withdraw the shifter shafts through left side of main case and remove shifters from above.

Install the shifter shafts and shifters by reversing the removal procedure and on series A (648000 and up), make certain that flat on right end of shifter shafts fully engages the shifter shaft lock plate (7—Fig. JD117). To align the shifter gates, place shifter (13—Fig. JD116 or 10—Fig. JD117) in neutral position and turn adjusting screws (2) either way as required to align the

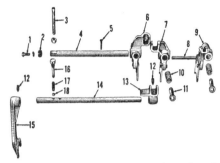

Fig. JD116—Exploded view of series A prior 648000 and series B prior 201000 shifter shafts and shifters.

1. Lock screw	10. Pawl spring
2. Adjusting screw	11. Shifter pawl
3. Set screw	12. Set screw
4. Shifter shaft	13. Shifter arm
5. Cotter pin	14. Fifth-sixth shifter shaft
6. First, third and reverse shifter	15. Fifth-sixth shifter
7. Second and fourth shifter	16. Detent plug
8. Lock out pin	17. Detent spring
9. Overdrive shifter	18. Detent ball (plunger)

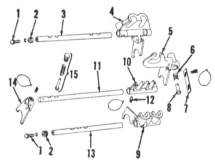

Fig. JD117—Exploded view of series A 648000 and up shifter shafts and shifters.

1. Lock screw	10. Fourth and sixth speed shifter
2. Adjusting screw	11. Fourth and sixth shifter shaft
3. Shifter shaft	12. Set screw
4. Second and fifth speed shifter	13. First, third and reverse shifter shaft
5. Overdrive shifter	14. Fourth and sixth speed shifter arm
6. Pawl spring	15. Detent assembly
7. Lock plate	
8. Pawl	
9. First, third and reverse shifter	

other shifter gates with the gate on shifter (10—Fig. JD117 or 13—Fig. JD116). For procedure for overhaul of the sliding gear shaft refer to paragraph 203.

Series B (after 200999)-G

197. To remove the transmission shifter shafts and shifters, first remove the transmission top cover as outlined in paragraph 191 and flywheel as outlined in paragraph 110: Remove the transmission shaft cover from **left** side of main case and remove the belt pulley and clutch as outlined in paragraph 177. On B, remove cover (30—Fig. JD127) from reduction gear case and the nut from the **right** end of the sliding gear shaft drive shaft (D) then remove the reduction gear main cover from **right** side of main case. On series B, remove nut and gear from left end of sliding gear shaft. Move each shifter along the shifter shafts until the shifter pawls rise. Install a cotter pin or nail in the hole in each shifter pawl to lock the pawls in the raised position. Remove the detent plunger spring and plunger from the vertical drilled hole which is located to the left of transmission top cover opening and on series G remove the vertical lock screw (3—Fig. JD121). Remove cap screws (1—Figs. JD120 & 121), adjusting screws (2) and pointed set screw (12). Withdraw shifter shafts from the main case and remove the shifters from above.

Install the shifter shafts and shifters by reversing the removal procedure and on series B, make certain that flat on right end of shifter shafts fully engages the shifter shaft lock plate (21). Adjust the position of the shifter shafts with adjusting screws (2) so that gears on sliding gear shaft mesh evenly with gears on the countershaft. On series B, place gears in

third speed position (with detent pawls in grooves) and turn the lock plate retaining screw (22) on the right side of main case until end of screw just contacts the edge of shifter, then back off the screw one full turn and lock the screw with the lock nut. For procedure for overhaul of the sliding gear shaft refer to paragraph 203.

Model D

200. Remove transmission cover as in paragraph 192. Remove cap screws attaching shift quadrant (27—Fig. JD122) to transmission case. Loosen

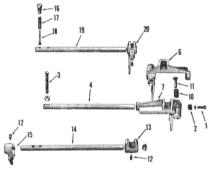

Fig. JD121—Exploded view of series G transmission shifter shafts and shifters.

1. Lock screw	13. Shifter arm
2. Adjusting screw	14. Shifter shaft
3. Set screw	15. Shifter
4. Shifter shaft	16. Detent plug
6. Shifter	17. Spring
7. Shifter	18. Plunger
10. Pawl spring	19. Underdrive shifter shaft
11. Pawl	20. Underdrive shifter
12. Set screw	

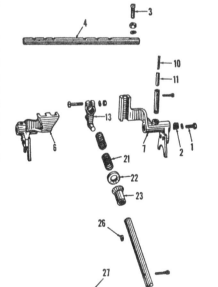

Fig. JD122—Exploded view of model D transmission shifter mechanism.

1. Lock screw	11. Stop pin
2. Adjusting screw	13. Shifter arm
3. Set screw	21. Lever spring
4. Shifter shaft	22. Cork ring
6. Second-third shifter	23. Shroud
7. Low-reverse shifter	24. Shift lever
10. Pin spring	26. Lever key

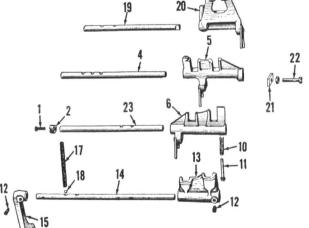

Fig. JD120—Exploded view of series B 201000 and up shifter shafts and shifters.

1. Lock screw	
2. Adjusting screw	
4. First-third-reverse shifter shaft	
5. First-third-reverse shifter	
6. Second-fifth shifter	
10. Pawl spring	
11. Pawl	
12. Set screw	
13. Fourth-sixth shifter	
14. Fourth-sixth shifter shaft	
15. Fourth-sixth fork	
17. Detent spring	
18. Detent ball	
19. Underdrive shifter shaft	
20. Underdrive shifter	
21. Lock plate	
22. Stop screw	
23. Second-fifth shifter shaft	

clamp bolt on shift lever arm, remove Woodruff key (26) and slide shift lever rearward. Note postion of springs (21), cork (22) and shroud (23) for reassembly.

Remove pulley and clutch as outlined in paragraph 177 and remove reduction gear cover.

Remove lock screw (1) and adjusting screw (2) from right end of shifter shaft. Loosen set screw (3) inside of transmission case and withdraw shifter shaft and shifters.

Reassemble in reverse order and adjust position of shifter shaft with adjusting screw (2) so that gears on spline shaft (sliding gear shaft) mesh evenly with gears on differential. When reinstalling the reduction gear cover, make certain that cover is centered around the clutch operating sleeve. For procedure for overhaul of the sliding gear shaft refer to paragraph 204.

Series H

201. Remove the flywheel as outlined in paragraph 110 and transmission top cover as outlined in paragraph 193. Slide shifter (6—Fig. JD 123) along the shifter shaft until the pawl rises. Install a cotter pin or nail in the hole in the shifter pawl to lock the pawl in the raised position. Unbolt the shifter shaft bearing (21) from left side of main case and withdraw bearing and shaft through left side of main case and shifters from above.

Reassemble in reverse order and adjust position of shifter shaft with adjusting screw (2) so that gears on sliding gear shaft mesh evenly with gears on countershaft. For procedure for overhaul of the sliding gear shaft refer to paragraph 205.

SLIDING GEAR (SPLINE) SHAFT

Series A-B-G

203. Remove shifter shafts and shifters as outlined in paragraph 196 or 197.

On series A prior to 648000 and series B, remove nut from left end of sliding gear shaft.

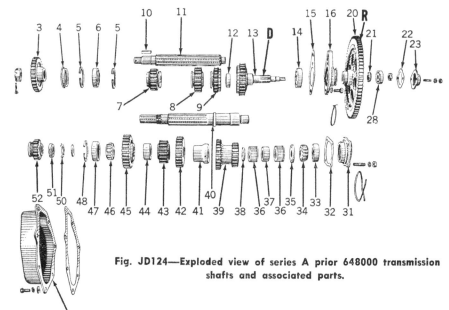

Fig. JD124—Exploded view of series A prior 648000 transmission shafts and associated parts.

3. Fifth and sixth drive gear	12. Bearing
4. Oil retainer	13. Sliding gear shaft drive gear
5. Snap ring	14. Bearing
6. Bearing	15. Gasket
7. First speed pinion	16. Bearing cover
8. Second speed pinion	20. First reduction gear
9. Sliding gear shaft drive pinion	21. Spacer
10. Key	22. Gasket
11. Sliding gear shaft	23. Cover
	28. Bearing

31. Bearing housing	43. Differential drive pinion
32. Shim gaskets	44. Spacer
33. Bearing cup	45. First speed gear
34. Cone	46. Bearing cone
35. Thrust washer	47. Cup
36. Idler bearing	48. Snap ring
37. & 38. Spacer	50. Locking washer
39. Idler gear	51. Locking nut
40. Countershaft	52. Fifth and sixth speed pinion
41. Spacer	
42. Second speed gear	

On all models remove gear, oil retainer and snap ring from left end of sliding gear shaft. Pull the shaft out through left side of main case while bumping the drive pinion against the bearing on right end of shaft to remove the bearing. Withdraw the shaft through the left side of case and remove gears from above.

Refer to the transmission exploded views (Figs. JD124, 125, 126, 127 or 129) and install the sliding gear shaft be reversing the removal procedure. Before reinstalling the shielded ball bearing to right hand end of sliding gear shaft pack the open side with wheel bearing grease. Always install this shielded bearing with the shield nearest to the left side of the tractor so as to retain the packed-in lubricant. For procedure for overhaul of the sliding gear shaft drive gear refer to paragraph 207.

Model D

204. Remove cover and shifter shaft and shifters as outlined in paragraph 200. Remove four cap screws retaining bearing quill (17—Fig. JD 130) to transmission case. Remove left cover, snap ring (3) from left end of shaft and remove left bearing (4) by bumping it with high and intermediate sliding gear (7). Withdraw shaft through opening in right side of case. Press reduction gear off shaft and remove bearing and quill. Install the sliding gear shaft by reversing the removal procedure. For procedure for overhauling the reverse gear shaft refer to paragraph 213.

Series H

205. Remove cover and shifter shaft and shifters as outlined in paragraph 201. Remove pulley and clutch as outlined in paragraph 180 and remove reduction gear cover and gear.

Remove cap screws retaining sliding gear shaft **right** bearing housing (8—Fig. JD131) to case and remove housing. Withdraw shaft assembly through opening in right side of case. If equipped with power shaft (P.T.O.), remove left bearing housing (1) and remove nut, bearing and bevel gear from left end of shaft before removing shaft.

Fig. JD123 — Exploded view of series H transmission shifter shaft and shifters. Item (21) is a shaft bearing.

1. Lock screw	6. Shifter
2. Adjusting screw	7. Shifter
3. Set screw	10. Pawl spring
4. Shifter shaft	11. Shifter pawl

Reassemble in reverse order and vary thickness of gaskets and shims (7) between right bearing housing and case to remove all bearing play but permitting shaft to turn freely. Gaskets and shims (2) between left bearing housing and case control position of bevel gear in relation to bevel gear (not shown) on power shaft. Refer to Power Shaft, paragraph 222 for gear mesh adjustment procedure.

When reinstalling reduction gear cover, make certain that the reduction gear cover is centered around the clutch operating sleeve. For procedure for overhauling the countershaft refer to paragraph 211.

SLIDING GEAR SHAFT DRIVE GEAR

(This is part "D" in exploded views of transmission.)

Series A-B-G

207. Remove the sliding gear shaft as outlined in paragraph 203.

On series B prior to 201000 and on all series A tractors remove belt pulley and clutch as outlined in paragraph 177, right brake and reduction gear cover.

Remove the reduction gear (R—Figs. JD124, 125, 126, 127 & 129) from right end of the sliding gear shaft drive gear (D). Remove bearing retainer (cover) from right side of main case and remove the drive gear through transmission cover opening.

Refer to the transmission exploded views and install the drive gear by reversing the removal procedure. For procedure for overhauling the countershaft refer to paragraph 209.

COUNTERSHAFT

(Countershaft is part (40) in Fig. JD124, (33) in JD125, (43) in JD126, (45) in JD127, (43) in JD129 and (22) in JD131.)

Series A-B-G

209. Remove sliding gear shaft and sliding gear shaft drive gear as in paragraphs 203 and 207. Remove countershaft right bearing housing

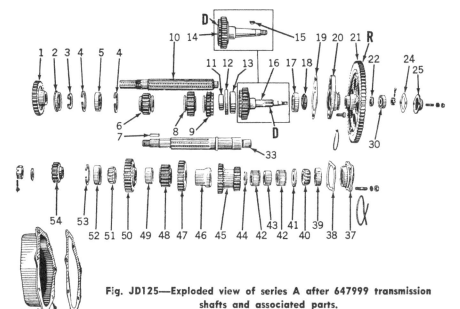

Fig. JD125—Exploded view of series A after 647999 transmission shafts and associated parts.

1. Fourth and sixth speed gear
2. Oil retainer
3. & 4. Snap ring
5. Bearing
6. First and third speed pinion
7. Key
8. Second and fifth speed pinion
9. Sliding gear shaft drive pinion
10. Sliding gear shaft
11. Bearing inner race
12. Snap ring
13. Bearing rollers
14. Sliding gear shaft drive gear (648000-657392)
15. Woodruff key
16. Sliding gear shaft drive gear (657393 and up)
17. Bearing
18. Oil collar
19. Gasket
20. Cover
21. First reduction gear
22. Spacer
24. Gasket
25. Cover
30. Bearing
33. Countershaft
37. Bearing housing
38. Shim gasket
39. Cup
40. Cone
41. Thrust washer
42. Idler bearing
43. & 46. Spacer
44. Collar
45. Idler gear
47. Second and fifth speed gear
48. Differential drive pinion
49. Spacer
50. First and third speed gear
51. Cone
52. Cup
53. Snap ring
54. Fourth and sixth speed pinion
60. Cover

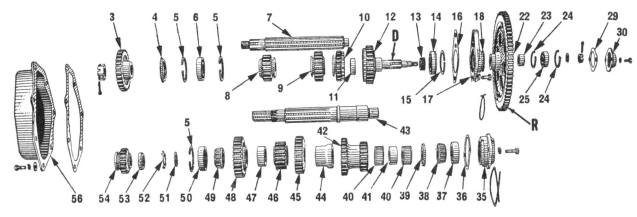

Fig. JD126—Exploded view of series B prior 201000 transmission shafts and gears.

3. Fifth and sixth speed gear
4. Oil retainer
6. Bearing
7. Sliding gear shaft
8. First, third and reverse pinion
9. Second and fourth pinion
10. Sliding gear shaft drive pinion
11. Bearing
12. Sliding gear shaft drive gear
13. Collar
14. Bearing
15. Snap ring
16. Gasket
17. Cover
18. Retainer
22. First reduction gear
23. Spacer
24. Snap ring
25. Bearing
29. Gasket
30. Cover
35. Bearing housing
36. Shim gaskets
37. Cup
38. Cone
39. Thrust washer
40. Roller bearing
41. & 44. Spacer
42. Idler gear
43. Countershaft
45. Second and fourth speed gear
46. Differential drive pinion
47. Spacer
48. First and third speed gear
49. Cone
50. Cup
51. Washer
52. Locking washer
53. Locking nut
54. Fifth and sixth sliding pinion

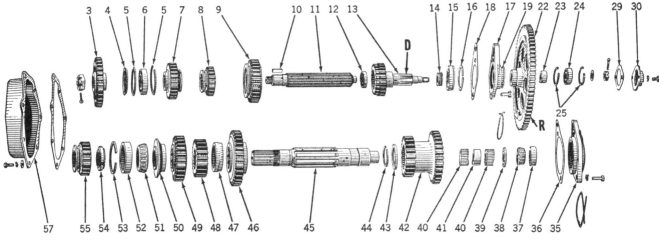

Fig. JD127—Exploded view of series B 201000 and up transmission shafts and gears.

3. Fourth & sixth speed gear
4. Oil retainer
5. Snap ring
6. Bearing
7. Second & fifth speed pinion
8. First & third speed pinion
9. Sliding gear shaft drive pinion
10. Key
11. Sliding gear shaft
12. Bearing
13. Sliding gear shaft drive gear
14. Oil collar
15. Bearing
16. & 25. Snap ring
17. Cover
18. & 29. Gasket
19. Retainer
22. First reduction gear
23. Spacer
24. Bearing
30. Cover
35. Bearing housing
36. Gasket
37. Cup
38. Cone
39. Thrust washer
40. Roller bearing
41. & 47. Spacer
42. Idler gear
43. Collar
44. & 53. Snap ring
45. Countershaft
46. First & third speed gear
48. Differential drive pinion
49. Second & fifth speed gear
50. Spacer
51. Cone
52. Cup
54. Lock nut
55. Fourth & sixth speed pinion
57. Cover

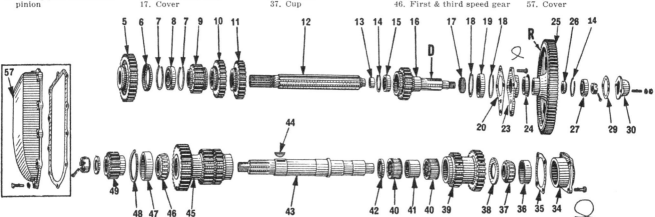

Fig. JD129—Exploded view of series G transmission shafts and gears.

5. Fifth & sixth speed pinion
6. Oil retainer
7. Snap ring
8. Bearing
9. First & second speed pinion
10. Third & fourth pinion
11. Sliding gear shaft drive pinion
12. Sliding gear shaft
13. Bearing race
14. Snap ring
15. Bearing
16. Sliding gear shaft drive gear and shaft
17. Collar
18. Snap ring
19. Bearing
20. Gasket
23. Bearing cover
24. Retainer
25. First reduction gear
26. Spacer
27. Bearing
29. Gasket
30. Cover
34. Bearing housing
35. Shim gaskets
36. Cup
37. Cone
38. Thrust washer
39. Idler gear
40. Idler bearings
41. Spacer
42. Washer
43. Countershaft
44. Key
45. Cluster gear
46. Cone
47. Cup
48. Snap ring
49. Fifth & sixth speed gear
57. Cover

Remove nut and/or gear from left end of countershaft. Pull shaft through case and withdraw gears and spacers from above.

Reassemble in reverse order and on models so equipped, tighten the nut securely on left end of shaft. Vary thickness of shims and gaskets between right bearing housing and case to provide 0.001-0.004 end play for the countershaft. For procedure for overhauling the power shaft used to drive the PTO system refer to paragraph 215.

Series H

211. Remove sliding gear shaft as described in paragraph 205. Remove both axle and housing assemblies as outlined in paragraph 269 then remove rear cover and lift out the differential.

Remove screw and lock plate (23—Fig. JD131) retaining right end of countershaft to case. Withdraw shaft through right side of case and gears, washers and spacers through rear opening. Gears can be pressed off pinion after removing snap rings.

Reassemble and reinstall in reverse order. There should be no end play between gears and pinion. Select gear retaining snap rings (12) of a thickness that will retain gears rigidly on pinion. Be sure a thrust washer (11) is installed at both ends of gears assem-

3. Snap ring
4. Left bearing
5. Gear key
6. Sliding gear shaft
7. Second-third sliding gear
8. First-reverse sliding gear
15. Snap gear
16. Bearing
17. Bearing quill
19. Reduction gear
24. Side cover
25. Reverse gear shaft
38. Reverse gear
40. Gear bushing
42. Adjusting washer
47. Dowel pin
48. Snap ring

Fig. JD130—Exploded view of model D transmission shafts and associated parts.

bly. For procedure for overhaul of
the reverse gear shaft refer to paragraph 214.

REVERSE GEAR AND SHAFT

Model D

213. Remove spline shaft (sliding
gear shaft) as described in paragraph
204.

Withdraw reverse gear shaft (25—
Fig. JD130) through right side of
transmission case and gear and thrust
washers from above. Bushing (40) in
gear is renewable. Press new bushing in and hone or ream, if necessary,
to obtain a free running fit. Reinstall
in reverse order; be sure thrust washers are in place. For procedure for
overhauling the power shaft used to
drive the PTO system refer to paragraph 221.

Series H

214. Remove sliding gear shaft as in
paragraph 205. Reverse shaft (24—
Fig. JD131) is retained by sliding gear
shaft right bearing housing (8) which
will be removed in getting the sliding
gear shaft out of the transmission case.
Withdraw shaft through right side of
case and gear from above. Bushings

(25) in gear are renewable. If new
bushings are installed in gear, hone
or ream, if necessary, to obtain a free

running fit. For procedure for overhauling the power shaft used to drive
the PTO system refer to paragraph 222.

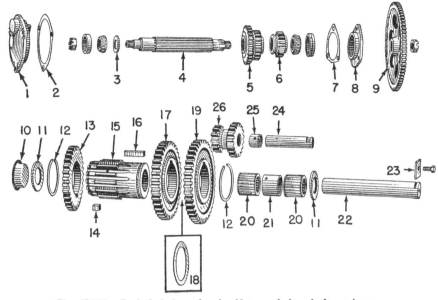

Fig. JD131—Exploded view of series H transmission shafts and gears.

1. Left bearing housing
2. Shim gasket
3. Thrust washer
4. Sliding gear shaft
5. Second-third sliding gear
6. First-reverse gear
7. Shim gasket
8. Right bearing housing
9. Reduction gear
10. Cover
11. Thrust washer
12. Snap ring
13. Third gear
14. Third gear key (after 47795)
15. Main drive pinion
16. Gear key (after 47795)
17. Second gear
18. Spacer (prior to 2211)
19. First gear
20. Bearing
21. Bearing spacer
22. Countershaft
23. Lock plate
24. Reverse gear shaft
25. Gear bushing (2)
26. Reverse gear

POWER SHAFT (P.T.O. SYSTEM)

Series A-B-G

215. **R&R AND OVERHAUL.** Remove transmission cover and countershaft as outlined in paragraphs 190 or
191 and 209.

Remove shifter arm (35—Fig. JD140
or 46—Fig. JD141) from shifter lever.
Remove shifter shaft (30 or 43) and
shifter (31 or 48), catching ball and
spring (33 or 45) as they fly out. Remove right bearing cover (26 or 39),
pull driving shaft and bearing to right
side of case and pull bearing off shaft.
Lift shaft and gears upward towards
left side of case and remove from
above. Shaft left bushing (21—Fig.
JD140) on A and B series can be driven out of case. Hone or ream new
bushing, if necessary, to provide a
free running fit. Series G is fitted with
a roller bearing (30—Fig. JD141) retained in the case by snap rings (29).

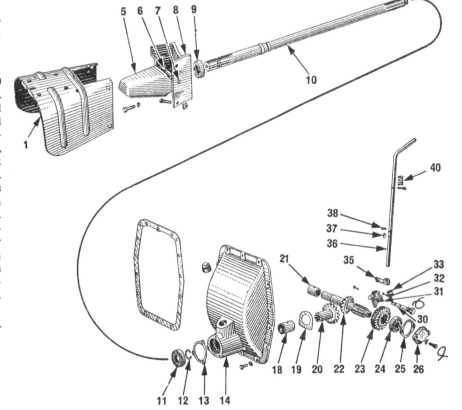

1. P.T.O. shield
5. Guard
6. Guard spring
7. Pin
8. Oil seal housing
9. Oil seal
10. P.T.O. shaft
11. Bearing
12. Snap ring
13. Gasket
18. Bushing
19. Thrust washer
20. Bevel gear
21. Bushing
22. Bevel pinion & shaft
23. Sliding gear
24. Bearing
25. Shim gasket
26. Bearing cover
30. Shifter shaft
31. Gear shifter
32. Ball
33. Spring
35. Shifter arm
36. Lever
37. Key
38. Snap ring
40. Spring

**Fig. JD140—Exploded view of early production series B power shaft and associated parts.
Series A and late production series B are similarly constructed. The differences are evident after an examination of the unit. Item (14) is called a basic housing when tractor
is equipped with a hydraulic lift.**

Remove rear cover or basic housing and pull the driven shaft out of transmission case. Remove snap ring, if tractor is so equipped, from the driven bevel gear and remove gear and shaft from front of boss. Bushing (18—Fig. JD140) in boss of A and B series transmission case is renewable and should be honed or reamed, if necessary, to provide a free running fit. Series G is fitted with bearings (25 & 27—Fig. JD141).

Reinstall in reverse order, make certain that thrust washer (19—Fig. JD 140 or 28—Fig. JD141) is properly located between gear and transmission boss and adjust the mesh position and backlash as outlined in the following paragraph.

220A. BEVEL GEARS ADJUSTMENT. The power shaft bevel gears should be meshed so that heels are in register and backlash is 0.006-0.010. On series A and B, the mesh position is controlled by thrust washer (19—Fig. JD140); on series G by shim gaskets (24—Fig. JD141). Backlash is controlled by shim gaskets (25 or 24).

Model D

221. R&R AND OVERHAUL. Remove cap screws retaining power shaft rear bearing quill (10—Fig. JD142) to transmission case and withdraw driven shaft (63) and quill as a unit. Remove cap screws retaining power shaft housing (38) to left side of transmission case and remove entire unit from left side of case. To remove drive spur gear (44) from sliding gear shaft (59), remove shifter then slide gear off shaft.

To disassemble the unit, remove nut from inner end of spindle (36). Remove cap screws retaining spindle to housing, withdraw spindle and remove gear (43). Remove cap screw and lock plate (25) from bevel driven gear bearing adjusting nut (26) and remove adjusting nut and gear.

Reassemble in reverse order. By means of nut (26) adjust bevel driven gear bearings to provide 0.001-0.004 end play for bevel gear (41). Adjust idler (bevel drive gear) gear bearings to provide 0.001-0.004 end play for idler gear (43) by means of nut on inner end of spindle (36). Gear backlash and mesh is controlled by thickness of shims (34) between outer end

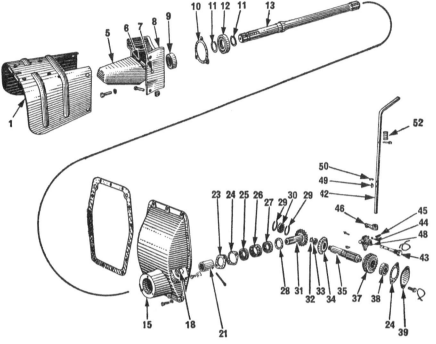

Fig. JD141—Exploded view of series G power shaft and associated parts. Item (15) is called a basic housing when tractor is equipped with a hydraulic lift.

1. P.T.O. shield	13. P.T.O. shaft	30. Bearing	42. Lever
5. Guard	18. Plug	31. Pinon and shaft	43. Shifter shaft
6. Guard spring	21. Coupling	32. Snap ring	44. Ball
7. Pin	23. Cover	33. Bearing race	45. Spring
8. Oil seal housing	24. Shim gasket	34. Bevel gear	46. Shifter arm
9. Oil seal	25 & 27. Bearing	35. Shaft	48. Shifter
10. Gasket	26. Spacer	37. Sliding gear	49. Key
11. Snap ring	28. Thrust washer	38. Bearing	50. Snap ring
12. Bearing	29. Snap ring	39. Cover	52. Spring

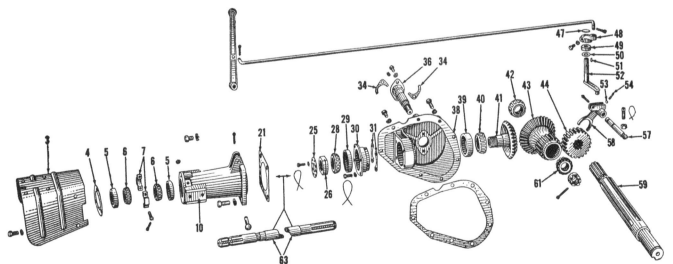

Fig. JD142—Exploded view of model D power shaft and associated parts.

4. Shims	25. Lock washer	34. Shims	42. Cone	49. Oil seal	54. Spring
5. & 6. Bearing	26. Adjusting nut	36. Spindle	43. Idler gear	50. Washer	58. Shift fork
7. Bearing collar	28. & 29. Bearing	38. Housing	44. Drive gear	51. Woodruff key	59. Sliding gear shaft
10. Quill	30. Bearing cage	39. & 40. Bearing	47. Welch plug	52. Shift arm	61. Bearing cone
21. Gasket	31. Shims	41. Bevel gear	48. Shift lever	53. Ball	63. P.T.O. shaft

of idler (bevel drive) gear spindle and housing and by shims (31) between bevel driven gear bearing cage and housing. I&T recommended backlash is 0.006-0.010. Power shaft rear bearings in rear quill are adjusted by shims (4) between rear bearing cover and rear quill.

Series H

222. R&R AND OVERHAUL. Remove shaft rear bearing housing (4—Fig. JD143) and withdraw driven shaft (6), bearings and gear (9) as a unit. Remove sliding gear shaft left bearing cover (11) and remove nut, bearing, thrust washer (13) and gear (15) from end of shaft. Bushing (14) can be pressed out of gear. New bushing should be honed or reamed, if necessary, to a free fit.

Reassemble in reverse order. Bevel gears should be meshed so that heels are in register and backlash is from 0.006-0.010. Sliding gear shaft bearing end play is controlled by shims (7—Fig. JD131) located under bearing housing (8) which is on right side of main case. Gear mesh adjustment is controlled by shims and gaskets (12—Fig. JD143) between sliding gear shaft bearing housing and transmission case. Gear mesh should be adjusted **after** the bearing end play has been adjusted. To move driving gear (15) into deeper mesh with driven gear (9) remove a shim or shims (12) and install the same shim or shims between sliding gear shaft right bearing housing (8—Fig. JD131) and transmission case. Reverse this procedure if gear driving bevel gear is to be moved away from the driven gear. Bevel gear backlash is controlled by shim gaskets (5—Fig. JD143) between rear bearing housing and rear cover plate (10).

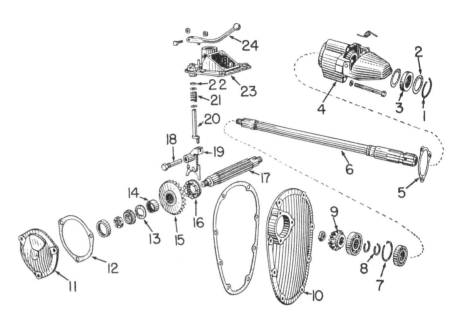

Fig. JD143—Exploded view of series H power shaft assembly.

1. Snap ring	7. Snap ring	12. Gear mesh shim gasket	18. Shifter shaft
2. Seal washer	8. Snap ring (1 prior to H47796)	13. Thrust washer	19. Shifter
3. Felt seal	9. Bevel gear	14. Gear bushing	20. Shifter crank
4. Housing & flipper guard	10. Rear cover	15. Driving gear	21. Spring
5. Gear backlash shim gasket	11. Transmission sliding gear shaft left cover	16. Shifting collar	22. Felt seal
6. Shaft		17. Transmission sliding gear shaft	23. Transmission cover
			24. Shift lever

DIFFERENTIAL, FINAL DRIVE AND REAR AXLE

DIFFERENTIAL

Series A-B-G

235. REMOVE AND REINSTALL. On ALL models remove rear axle assembly as in paragraph 266 or 267, and remove both brake assemblies. Remove differential left bearing quill (Figs. JD145, 146 or 148).

235A. On series B 201000-up perform work outlined in paragraph 235 and remove pulley and clutch as outlined in paragraph 177 and remove reduction gear cover. Move differential right bearing quill (14—Fig. JD146) out approximately one inch after loosening its retaining screws.

235B. On High Crop models, perform work outlined in paragraph 235 and disengage the inner snap ring (9—Fig. JD148) and slide the snap ring and sprocket in and against the differential side bevel gears.

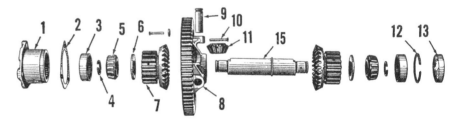

Fig. JD145—Exploded view of series A differential. Series B prior 201000 and series G are similar. Snap rings (4) are not used on series G.

1. Left bearing quill	5. Bearing cone	8. Spider with gear	12. Snap ring
2. Shim gasket	6. Thrust washer	9. Pinion shaft	13. Bearing cover
3. Bearing cup	7. Side gear and bull pinion	10. Pinion shaft rivet	14. Right bearing quill
4. Snap ring (not on G)		11. Pinion gear	15. Differential shaft

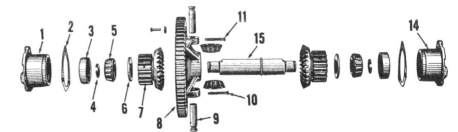

Fig. JD146—Exploded view of series B 201000 and up differential. See Fig. JD145 for legend except pinion in this view is not called out.

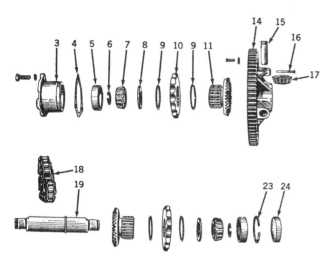

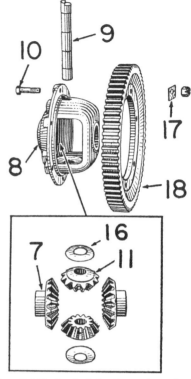

Fig. JD148—Exploded view of models AH and GH high crop differential.

3. Bearing quill
4. Shims
5. Cup
6. Snap ring
7. Cone
8. Thrust washer
9. Snap ring
10. Sprocket
11. Differential bevel gear
14. Differential gear and spider
15. Pinion shaft
16. Rivet
17. Pinion
18. Drive chain
19. Differential shaft
23. Snap ring
24. Bearing cover

Fig. JD150—Exploded view of series H differential.

7. Side gear
8. Differential case
9. Pinion shaft
10. Gear retaining bolt (6)
11. Pinion gear
16. Thrust washer
17. Retaining washer
18. Main drive gear

236. **On all models,** slide differential assembly to the left side of the transmission case, lift right end and move assembly forward or rearward as required so that assembly can be removed through rear opening.

Reinstall in reverse order and vary number of shims (2—Figs. JD145 and 146 or 4—Fig. JD148) between bearing quill and transmission case to provide 0.001-0.004 end play for the differential assembly.

237. **OVERHAUL.** Remove snap rings (4—Figs. JD 145 and 146 or 6—Fig. JD148) retaining bearing cones and side gears to shaft. Using a puller, remove bearings, gears and thrust washers. Remove rivets retaining bevel pinion shafts to spider and remove pinion shafts and pinions. Reassemble in reverse order. Select proper thickness of snap rings (4) on the A and B series to retain bearing cones firmly against shoulder of shaft. Thoroughly lubricate pinions and shafts before assembly.

Model D

238. **R&R AND OVERHAUL.** Remove both rear axle housing (quill) assemblies as in paragraph 268. Support differential in chain hoist. Remove differential left quill and left bearing and thrust washer. Remove

right quill. Move assembly towards left side of transmission case and lift while moving toward rear of case. Pull top of assembly towards right side of case and continue to lift while guiding left end of shaft through notch provided in left side of case. Reinstall in reverse order. Thrust washers (16—Fig. JD149) in quills determine end play of the unit which should be approximately 1/16 inch. Adjusting washers (11) should be added, if needed, to reduce end play of side gears (10) to approximately 1/32 inch.

Series H

239. **REMOVE AND REINSTALL.** Remove rear cover and remove rear axle housing assemblies as outlined in paragraph 269. Withdraw differential assembly through rear opening. Reinstall in reverse order.

240. **OVERHAUL.** Remove bolts (10—Fig. JD150) and nuts retaining spur ring gear (18) to differential case (8) and remove gear. Differential pinion shaft (9) can be withdrawn from case, and pinion and side gears removed. Reassemble in reverse order, making certain that thrust washers (16) are in place between pinion gears and case and pinion shaft (9) is locked in place by retaining washer (17).

REAR WHEELS

Series A (except High Crop)-B-G (except High Crop)-H-Model D

245. To remove rear wheels, remove cap screws retaining hub clamp to hub. Lubricate holes in hub clamp and install cap screws in removing holes as in Fig. JD151. Turn cap screws progressively to remove hub clamp. When reinstalling clamp, tighten cap screws progressively to prevent damaging hub or clamp.

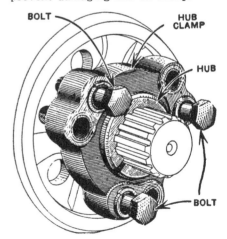

Fig. JD151—Removing rear wheel hub clamp from series A-B-G and H. Clamp retaining bolts are removed, then installed in holes in clamp so that inner ends contact hub. Turning bolts in removes clamp.

1. Differential shaft
2. Dowel pin (2)
3. Spider bolt (9)
4. Second gear
5. Pinion shaft (3)
6. Spider and third gear
7. First gear
8. Pinion thrust washer (3)
9. Pinion gear (3)
10. Side gear and sprocket
11. Adjusting washer (as required)
12. Thrust washer (2)
13. Inner race
14. Bearing
15. Outer washer
16. End thrust washer
17. Snap ring (2)
18. Outer race (2)
19. Bearing quill (2)

Fig. JD149—Exploded view of model D differential assembly.

Models AH-GH (High Crop)

247. To remove the rear wheels support rear of tractor and remove the stud nuts which retain the wheel hub to the stub axle shaft.

ADJUST AXLE BEARING
Series A (except High Crop)-B-G (except High Crop)-Model D

250. Support rear of tractor and remove rear cover or basic housing (top cover on model D). Remove cotter pin from inner end of axle shaft and tighten nut (2—Fig. JD152 or 154) to obtain a recommended axle end play of 0.001-0.004.

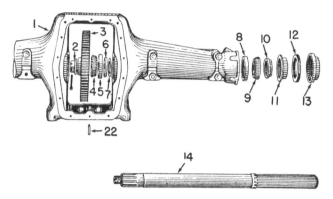

Models AH-GH (High Crop)

251. Support rear of tractor and remove cover (2—Fig. JD156) from the final drive gear housing. Remove cotter pin from inner end of the stub axle shaft and tighten the nut (5) to pre-load the bearings equivalent to 0.002 deflection of the housing when measured at the inner bearing boss.

Series H

252. Support rear of tractor and remove wheel. Bearings should be adjusted to 0.004-0.005 preload by means of shims (23—Fig. JD157). One method of obtaining this amount of preload is as follows: Loosen cap screws retaining axle housing (1—Fig. JD157) to transmission case and retighten the top cap screw. Vary the thickness of

shims (23) between outer bearing cover and axle housing to cause axle housing to separate 0.005 from transmission case. Measure this separation with a feeler inserted at about middle of front and rear of housing. Retighten balance of cap screws.

R&R WHEEL AXLE SHAFT, BULL GEAR, BEARINGS AND/OR SEALS
Series A (except High Crop)-B-G (except High Crop)

253. Support rear of tractor and remove rear cover or power lift and rear wheel. Loosen nut (2—Fig. JD152) on

Fig. JD152—Exploded view of series A rear axle assembly. Series B and G are similarly constructed.

1. Housing
2. Gear retaining nut
3. Final drive gear (bull gear) (2)
4. Inner cone
5. Inner cup
6. Inner oil seal
7. Oil seal washer
8. Outer cup
9. Outer cone
10. Spacer
11. Felt retainer (inner)
12. Felt seal
13. Felt retainer
14. Axle shaft
22. Dowel pin

Fig. JD155 — Model D tractor with right rear wheel, clutch, and reduction gear cover removed. Arrow on side of main case indicates direction to turn final drive eccentric quill to tighten main drive chain.

inner end of axle shaft. Drive a long tapered wedge between inner ends of axle shafts to loosen axle from bull gear as shown in Fig. JD158. Remove nut and withdraw axle shaft. The bull gear (27) can be renewed at this time. Outer bearing cup can be pulled with a puller. To remove inner bearing cup it is usually necessary to first remove the other wheel axle shaft. Reassemble in reverse order.

Be sure protective washer (7—Fig. JD152) is installed between oil seal and flange in housing and carefully install inner oil seal (6) with sharp edge of seal towards gear using a large plate and long bolt to pull seal in squarely.

On series A 584000 up, series B 201000 up, series G 13751 up, felt seal (12) can be removed without removing axle shaft by removing wheel and felt retainer from outer end of housing.

Refer to paragraph 250 for bearing adjustment.

Models AH-GH (High Crop)

254. Support rear of tractor and remove wheel, inner bearing cover (2—

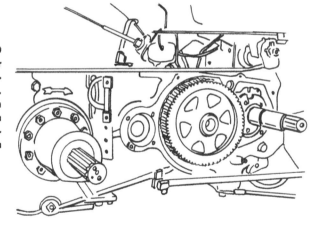

Fig. JD156) and final drive housing lower cover (30). Remove nut (5), bearing cone and snap ring (33) from inner end of wheel axle shaft. Drive the axle shaft out of final drive housing and bull gear. The bull gear (32) can be renewed at this time. The need and procedure for further disassembly is evident after an examination of the unit. Refer to paragraph 251 for bearing adjustment.

Model D

255. To remove bull sprocket follow the procedure outlined in paragraph 268. To renew wheel axle shafts or bearings perform the work outlined in paragraph 268, then bump shaft out of

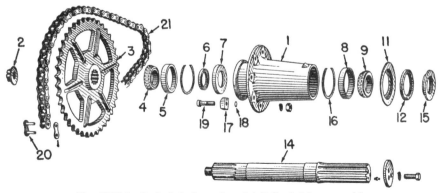

Fig. JD154—Exploded view of model D final drive assembly.

1. Axle quill	5. Inner cup	11. Felt retainer	17. Quill clamp (lug)
2. Sprocket retaining nut	6. Inner oil seal	12. Felt seal	18. Vellumoid washer
3. Sprocket	7. Oil seal retainer	14. Axle shaft	19. Quill clamp bolt
4. Inner cone	8. Outer cup	15. Spacer	20. Connecting link
	9. Outer cone	16. Snap ring	21. Chain

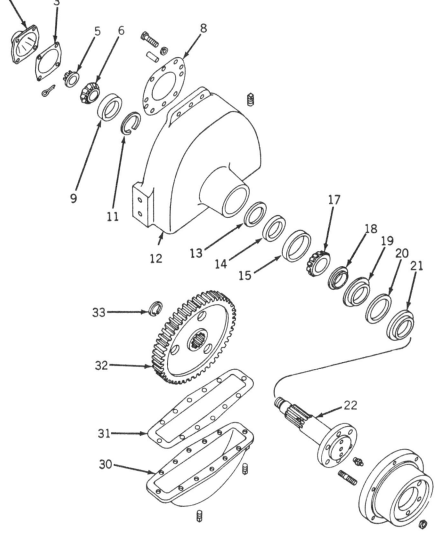

Models AH-GH (High Crop)

260. To remove the drive chain and/or driven sprocket, remove platform and rear cover or basic housing. Support rear of tractor and turn rear wheels until chain "master link" is accessible. Remove the master link and withdraw the chain. Remove nut and washer from inner end of drive shaft (13—Fig. JD160) and remove the driven sprocket.

Reinstall in reverse order and make certain that the drive chain "master link" is installed with open end toward center of tractor, and check and adjust the chain slack as per the following paragraph.

NOTE: To remove the drive sprocket, which is mounted on the differential side bevel gears, it is necessary to remove the differential as outlined in paragraph 235.

260A. ADJUST CHAINS. To check and adjust the slack in the drive chains, remove the inspection covers from case. The chains should have not less than ½-inch or not more than 1¾ inch slack. To adjust the slack, remove the brake drum and the brake housing retaining stud nuts. Slide the brake housing just off the studs and rotate the brake housing one stud hole at a time, either way, until desired chain slack is obtained and reinstall the brake housing.

Fig. JD156—Exploded view of AH & GH high crop final drive housing, wheel axle and bull gear assembly.

2. Bearing cover	11. Snap ring	18. Spacer
3. Gasket	12. Housing	19. Felt inner
5. Nut	13. Washer	retainer
6. Cone	14. Oil seal	20. Felt washer
8. Gasket	15. Bearing cup	21. Felt outer retainer
9. Bearing cup	17. Cone	22. Wheel axle shaft
		30. Housing cover
		31. Gasket
		32. Final drive (bull) gear
		33. Snap ring

1. Axle housing	23. Bearing adjusting shims
4. Inner cone	24. Bearing cover
5. Inner cup (in transmission case)	25. Dust excluder
6. Inner oil seal (in transmission case)	28. Bearing spacer
8. & 9. Outer cup & cone	29. Spacer
10. Spacer	30. Differential spacer
11. Felt retainer	31. Brake pedal
12. Inner felt	32. Pedal spring
14. Wheel axle shaft	33. Pedal shaft
22. Outer felt retainer	34. Brake cam
	35. Adjusting screw

housing. Shaft can be pressed out of outer cone, and cups and oil seal can be driven out of housing. Refer to paragraph 250 for bearing adjustment.

Series H

256. Remove axle housing assembly as outlined in paragraph 269. The inner seal (6—Fig. JD157) can be removed from main case at this time. Make certain that oil drain hole in case is open and clean. Use a puller and remove brake drum from shaft. Remove outer bearing cover (24—Fig. JD157) and pull shaft from housing. Reinstall in reverse order and adjust bearings as outlined in paragraph 252. To prevent oil leakage past inner bearing spacer (28), apply a coat of shellac or equivalent to inner surface of spacer.

257. The bull gear is bolted to and removed with the differential.

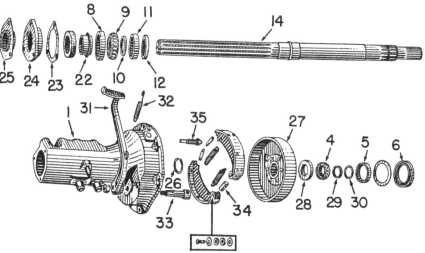

Fig. JD157—Exploded view of series H rear axle and brake assembly.

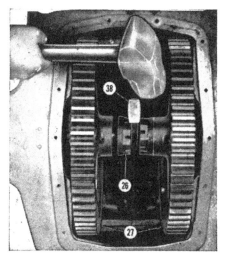

Fig. JD158—Removing series A, B and G bull gear (27), using a long taper wedge (38). The axle bearing adjusting nut is shown at (26).

After adjusting the chain slack, remove the brake shoe carrier from the brake housing and re-position the carrier so that brake pedal is in the proper position.

Model D

262. Support rear of tractor and remove top cover. Loosen the housing to transmission bolts (19—Fig. JD154). Remove cotter pins from connecting link (20) and remove link. Remove chain.

To remove sprocket (3), remove nut from inner end of axle shaft and force shaft out of sprocket. Reassemble in reverse order and adjust chain as in the following paragraph and bearings as in paragraph 250.

NOTE: To R&R the drive sprockets which are integral with the differential side bevel gears it is necessary to remove the differential as outlined in paragraph 238.

262A. ADJUST CHAINS. With quill to transmission clamp bolts partially tightened, rotate quill on transmission case in direction of arrow on side of case to tighten chain. Refer to Fig. JD155. Sprockets should turn freely with approximately one inch of chain slack at tightest position. Tighten quill retaining bolts securely.

R&R FINAL DRIVE GEAR HOUSING

Models AH-GH (High Crop)

264. To remove either final drive housing (12—Fig. JD156) support rear of tractor and remove wheel. Remove the housing retaining cap screws and remove the housing.

Install the housing by reversing the removal procedure and make certain that the housing locating dowels are in place before reconnecting the two housings.

R&R DRIVE SHAFT AND PINION
Models AH-GH (High Crop)

265. To remove the drive shaft (13—Fig. JD160), first remove the final drive gear housing as in paragraph 264 and the chain and driven sprocket as in paragraph 260. Remove bearing cone from inner end of drive shaft and withdraw shaft from housing.

Reassemble in reverse order and tighten nut (2) to obtain a drive shaft end play of 0.004.

R&R COMPLETE DRIVE SHAFT HOUSING ASSEMBLY

(This is item (1) in Fig. JD160 and all of its component parts as a unit)

Models AH-GH (High Crop)

266. Support rear of tractor and remove the drive chains as in paragraph 260. Unbolt the housing assembly from main case and remove the housing.

Install the housing assembly by reversing the removal procedure.

R&R COMPLETE AXLE HOUSING

(This is item (1) in Fig. JD152 and all of its component parts)

Series A (except High Crop)-B-G (except High Crop)

267. Support rear of tractor and remove platform. Remove cap screws retaining rear axle housing to transmission case and carefully separate housing from case. Housing is located by dowels in main case. Reinstall in reverse order making certain that gears mesh properly.

R&R LEFT OR RIGHT AXLE HOUSING

(This is item (1) in Figs. JD154 and 157 and its component parts)

Model D

268. To remove either axle housing assembly, support rear of tractor, block securely and remove wheel and fender. Remove top cover from main case. Remove nut from inner end of wheel axle shaft. Remove bolts (19—Fig. JD154) and clamps (17) retaining quill (axle housing) to transmission case and pull assembly out of case. The bull sprocket can be removed from the drive chain and lifted out of the tractor at this time.

Reassemble in reverse order with narrow section of quill (1) towards front of tractor. Install bolts and quill clamps (vellum washers (18) under lower bolts) and partially tighten bolts. Adjust the chain slack as outlined in paragraph 262A.

Series H

269. To remove either axle housing assembly, support rear of tractor and block securely. Remove cap screws retaining axle housing to transmission case and remove assembly. Reinstall in reverse order and adjust bearings as outlined in paragraph 252.

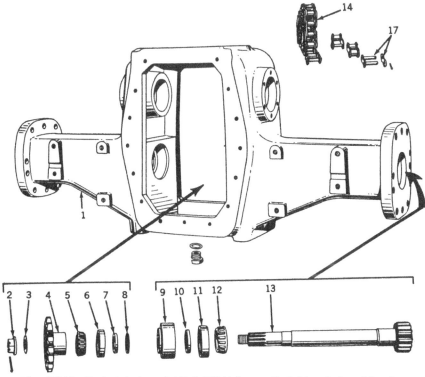

Fig. JD160—Exploded view of AH & GH high crop final drive shaft and housing.

1. Housing	5. & 6. Bearing
2. Nut	7. Inner oil seal
3. Washer	8. Washer
4. Sprocket	

9. Adapter	13. Drive shaft &
10. Outer oil seal	bull pinion
11. & 12. Bearing	14. Drive chain
	17. Master link

BRAKES

Series A-B-G

300. ADJUSTMENT. Tighten adjusting screw (35—Fig. JD161 or 162) or (23—Fig. JD163) which is located at top of brake housing until pedal free travel is reduced to approximately 3 inches.

301. R&R SHOES AND LINING. On all models remove platform if interference is encountered.

On A (except high crop) and B series, remove complete brake unit from tractor; then, remove nut from inner end of shaft (1—Fig. JD161) and remove gear (8 or 9) using a suitable puller or wedges. Loosen brake adjusting screw (35) and withdraw drum and shaft.

On G series (except high crop), remove nut retaining drum (27—Fig. JD162) to shaft, loosen brake adjusting nut and pull drum off shaft.

On high crop models, remove the nut (6—Fig. JD163) retaining the drum (8) to shaft (41), loosen the adjusting screw (23) and remove the drum.

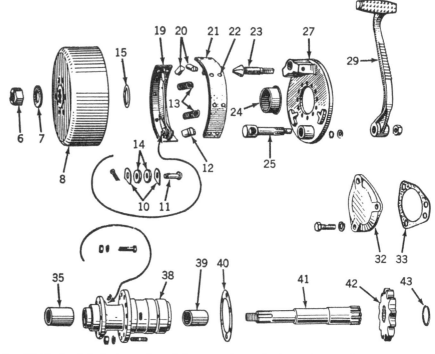

Fig. JD163—Exploded view of models AH and GH brake unit. Sprocket (42) meshes with the tractor drive chain.

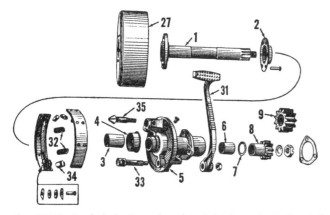

Fig. JD161—Exploded view of series A brake unit. Series B is similar except hub of gear (9) is flush with housing instead of extending into housing as with series A gear (8).

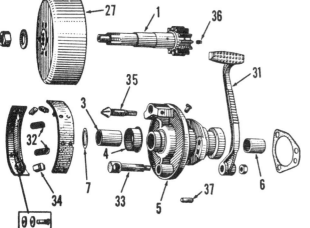

Fig. JD162—Exploded view of series G brake unit.

1. Brake drum shaft	
2. Dust guard	
3. Outer bushing	
4. Dust guard	
5. Brake housing	
6. Inner bushing	
7. Spacer washer	
8. Gear (A)	
9. Gear (B)	
27. Brake drum	
31. Pedal	
32. Spring	
33. Pedal shaft	
34. Brake cam	
35. Adjusting screw	

1. Brake drum shaft	
3. Outer bushing	
4. Dust guard	
5. Brake housing	
6. Inner bushing	
7. Thrust washer (2)	
27. Brake drum	
31. Pedal	
32. Spring	
33. Pedal shaft	
34. Brake cam	
35. Adjusting screw	
36. Shaft plug	
37. Dowel pin	

6. Nut	29. Pedal
7. Washer	32. Brake opening cover
8. Brake drum	
15. Washer	33. Gasket
19. Brake shoe	35. Outer bushing
20. Adjusting pins	38. Housing
21. Brake facing	39. Inner bushing
22. Rivet	40. Gasket
23. Adjusting screw	41. Brake shaft
24. Dust guard	42. Sprocket
25. Lever shaft	43. Snap ring
27. Brake shoe carrier	

On all models disconnect the shoe retracting springs and remove shoes from housing. Reassemble in reverse order and adjust as in paragraph 300.

302. REBUSH BRAKE SHAFTS. On A (except high crop) and B series, remove shoes as outlined in paragraph 301 and disassemble the unit. On other models, remove shoes as in paragraph 301, remove the housing unit from tractor and disassemble the unit.

NOTE: on high crop models, it is necessary to remove rear cover or basic housing and remove sprocket (42 —Fig. JD163) before removing the brake unit.

The brake shaft bushings can be driven out of housing and new ones pressed in. New bushings should be honed or reamed, if necessary, to provide a free fit. Reassemble in reverse order. Select spacer washers or thrust washers (7 or 15) to limit shaft end play to approximately 0.030. The riveted-on A and B drums are available separately from the shaft.

Model D Brake

303. ADJUSTMENT. Tighten adjusting nut (15—JD165) at lower end of band until pedal free travel is reduced to the desired amount. Check for clearance between lining and drum.

304. R&R BAND. Remove adjusting nut (15) and pin attaching band to crank (14). Reassemble in reverse order and adjust as in paragraph 303.

305. OVERHAUL UNIT. Remove drum. Remove cap screws retaining brake quill (6) to transmission case, remove quill and withdraw shaft. Inner cone and gear and outer cone can be pressed off shaft and oil seal (8) driven out of quill. Reassemble and reinstall in reverse order. Vary thickness of shims and gaskets (7) between quill and transmission case to provide a shaft end play of 0.001-0.004.

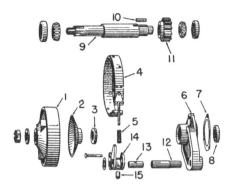

Fig. JD165—Exploded view of model D service brake.

1. Brake drum	8. Oil seal
2. Felt retainer	9. Brake shaft
3. Felt seal	10. Gear key
4. Brake band	11. Gear
5. Release spring	12. Pivot pin
6. Brake quill	13. Lever pin
7. Shim gasket	14. Crank

Series H Brakes

308. ADJUSTMENT. Tighten adjusting screw (35—Fig. JD157) located near top of each axle housing until pedal free travel is reduced to approximately 3 inches.

309. R&R SHOES AND LINING. Remove axle housing as in paragraph 269. Remove bearing spacer (28—Fig. JD 157) and drum with puller. Remove shoes. Reinstall in reverse order; adjust brakes as in paragraph 308 and refer to paragraph 252 for axle bearing adjustment. To prevent oil leakage past inner spacer, apply a coat of shellac or equivalent sealing compound to inner surface of spacer (28) during reassembly.

HYDRAULIC LIFT

The hydraulic system which was used on series H tractors and early production series A, B and G tractors was called a "Power Lift" system. The hydraulic system which is used on late production A, B and G tractors is called a "Powr-Trol" system. The differences between the "Power-Lift" and "Powr-Trol" systems are evident after an examination of Figs. JD168 and 187.

NOTE: The maintenance of absolute cleanliness of all parts is of utmost im-portance in the operation and servicing of the hydraulic system. Of equal importance is the avoidance of nicks or burrs on any of the working parts.

"Power Lift"

(Used on early production series A, B and G, and all series H tractors)

Series A-B-G

315. R&R ASSEMBLY. Detach implement, remove cap screws retaining assembly to transmission and remove assembly, carefully disengaging pump shaft from power shaft coupling. Reverse removal procedure to reinstall.

316. RENEW PISTON CUP. Drain power lift oil and remove valve housing (control box) (67—Fig. JD168) from rear of lift. Partially withdraw rock shafts (27) by removing cap screw from arm (37) and withdraw piston (11) crank and connecting rod as a unit. Install new leather cup (9) on piston, soak in oil and reassemble in reverse order.

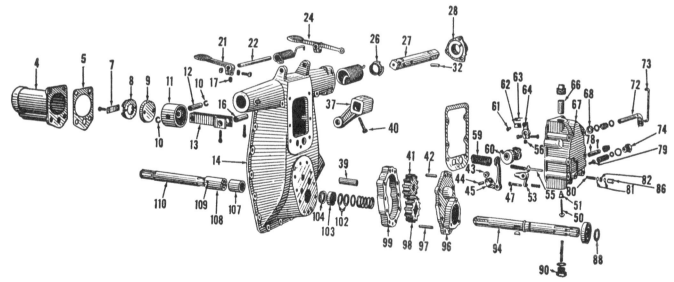

Fig. JD168—Exploded view of series G power lift. Series A and B are similarly constructed.

4. Cylinder	24. Trip pedal (right)	47. Cam trip pin	67. Valve housing	90. Screw plug
7. Lock plate	26. Shield	50. Check valve	68. Packing	94. Pump shaft
8. Cup retainer	27. Rock shaft (2)	51. Valve guide	72. Control shaft	96. Pump cover
9. Piston cup	28. Rock shaft bearing	53. Cam trip lever	73. Control rod	97. Dowel pin (lower)
10. Snap ring	32. Dowel pin	55. Throttle valve	74. Screw plug	98. Drive gear
11. Piston	37. Crank arm	56. Key washer	78. Cam follower shaft	99. Pump body
12. Piston pin	39. Idler gear shaft	60. Cam	79. Relief valve	102. Packing
13. Connecting rod	41. Idler gear	61. Pawl pin	80. Adjusting screw	103. Cup
14. Housing	42. Dowel pin (upper)	62. Pawl	81. Copper washer	104. Oil seal bearing
16. Crank pin	43. Follower roller	63. Pawl spring	82. Jam nut	107. Bushing
17. Key washer	44. Roller pin	64. Control lever	86. Cap nut	108. Coupling
21. Trip pedal (left)	45. Cam follower	66. By-pass valve	88. Snap ring	109. Rivet
22. Pedal shaft				110. Shaft

317. If lift does not hold implement in raised position and piston cup is in good condition, check for leaking valves and, if necessary, reseat or renew valves and seats. Check also leakage at gasket (5) between cylinder (4) and housing. Gasket renewal requires removal of power lift assembly and removal of cylinder. When installing gasket, apply a light coat of sealing compound and tighten cylinder retaining cap screws securely.

Series H

320. **R&R ASSEMBLY.** Remove magneto. Disconnect oil line, drain oil and remove cap screws retaining assembly to left side of governor case and remove assembly. Reinstall in reverse order and time magneto as outlined in paragraph 162.

321. **RENEW PISTON CUP.** Loosen set screw (2—Fig. JD170) on cylinder cover, remove piston retaining nut (6) and withdraw piston (9). Install new leather cup (16) on piston, soak in oil and reassemble in reverse order with arrows to top side. If oil leaks from cylinder assembly, inspect for dirt at plunger (12) and plunger retainer (11) in piston and clean thoroughly.

If lift does not hold in raised position, inspect the check valve (47—Fig. JD171) at lower end of the pump housing and reseat or renew the valve seating surfaces if necessary.

"Powr-Trol"

(Used on later production series A, B, and G.)

LUBRICATION AND BLEEDING
Series A-B-G

350. It is recommended that the "Powr-Trol" working fluid (same weight oil as used in the engine crankcase) be changed at least once-a-year. Drain housing and flush same with distillate. Fill the valve housing reservoir, start engine, operate the "Powr-Trol" lever several times to bleed the system and fill the remote control cylinder; then, refill the reservoir.

RENEW PISTON SEAL
Series A-B-G

350A. A general procedure for renewing the piston seal (9—Fig. JD180)

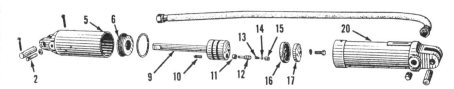

Fig. JD170—Exploded view of series H power lift cylinder.

2. Set screw	9. Piston	12. Plunger	15. Check
5. Cylinder cover	10. Spring	13. Spring	17. Cup retainer
6. Piston retainer	11. Plunger retainer	14. Ball	20. Cylinder

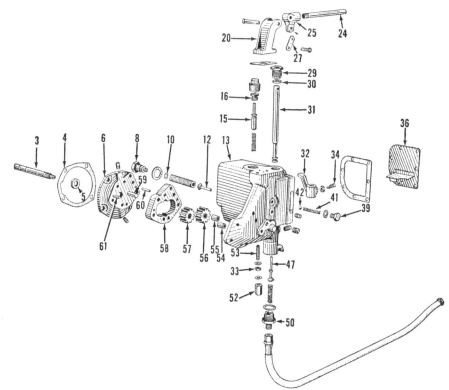

Fig. JD171—Exploded view of series H power lift pump and control unit.

3. Governor shaft	10. Adjusting washer (3)	24. Lever	33. Jam nut
4. Shim gasket	12. Relief valve	25. Lever head	36. Cover
5. Oil seal	13. Housing	27. Link (2)	39. Screw plug
6. Housing and pump cover	15. By-pass valve	29. Packing gland	42. Ball
8. Relief valve plug	16. Retainer plug	30. Packing	47. Check valve
	20. Lever bracket	31. Control valve	50. Valve cap
		32. Control	

is given in paragraph 316. To renew cylinder (4), it is necessary to remove basic housing (51) from tractor.

PUMP UNIT

Series A-B-G

351. **R&R AND OVERHAUL.** To remove the pump, first drain the hydraulic fluid and remove the power take-off shield. Remove the PTO external shaft, pump cover (45—Fig. JD180) retaining cap screws and disassemble the pump from the basic housing. Examine all parts and renew any which are excessively worn. When reinstalling the pump, make certain that the pump body is centered about the gears and tighten the cover retaining cap screws securely.

ADJUST ROCKSHAFT FAST DROP, SLOW RISE AND "POWR-TROL" METERING SCREW

Series A-B-G

353. To adjust the rockshaft "fast drop", remove cap nut, loosen jam nut and turn the throttle valve screw (73—Fig. JD183) "in" to increase or "out" to decrease speed of fast drop.

354. To adjust the rockshaft "slow rise", remove cap nut, loosen jam nut and turn the metering screw (83) "out" to decrease or "in" to increase speed of slow rise.

355. To adjust the "Powr-Trol" system metering screw so that it can be used for remote cylinder operation, remove cap nut, loosen jam nut and turn the metering screw (83) "in" until it seats firmly.

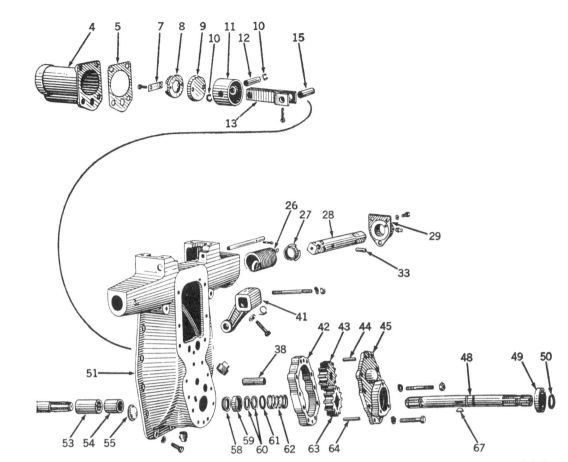

4. Cylinder
5. Gasket
7. Lock plate
8. Retainer
9. Piston cup
10. Snap ring
11. Piston
12. Piston pin
13. Connecting rod
15. Crank pin
26. Spring
27. Shield
28. Rockshaft
29. Rockshaft bearing
33. Dowel pin
38. Idler gear shaft
42. Pump body
43. Idler gear
44. & 64. Dowel pins
45. Pump cover
48. P.T.O. & pump shaft
49. Bearing
50. Snap ring
51. Basic housing
53. Coupling
54. Bushing
55. Inspection plug
58. Oil seal bearing
59. Cup
60. Packing
61. Oil seal washer
62. Spring
63. Pump drive gear

Fig. JD180—Exploded view of typical basic housing, hydraulic pump and rockshaft operating cylinder as used on series A, B and G in conjunction with the "Powr-Trol" valve housing assembly shown in Fig. JD187.

PRESSURE TEST

Series A-B-G

357. To check and adjust the relief valve cracking pressure, mount a pressure gage of sufficient capacity (at least 1500 psi) as shown in Fig. JD185.

Note: The shut-off valve must be located between the gage and the valve housing on the oil return line. With the shut-off valve open, move the "Powr-Trol" operating lever to raise position, allowing the working fluid to circulate through the tube and shut off valve and back to the valve housing. Close the valve slowly, and note the pressure reading as the relief valve opens. The relief valve cracking pressure should be 750-790 psi.

If the gage reading is not as specified, remove the relief valve spring plug (58—Fig. JD187) and add washers (60) to increase pressure; or remove washers to decrease pressure. Each washer represents approximately 35 psi.

If the specified gage pressure cannot be obtained, look for a failed or badly worn pump unit. Refer to paragraph 351 for overhaul of the pump.

CHECK VALVES

Series A-B-G

358. LEAK CHECK. Using a spare check valve screw plug, pressure gage, old inner tube valve stem and fittings, make up a leak detector as shown in Fig. JD189. Remove the upper check valve screw plug (69—Fig. JD187)

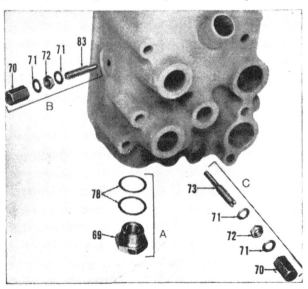

Fig. JD183—Series A, B and G "Powr-Trol" metering screw assembly (B), throttle valve screw assembly (C) and operating valve adjusting plug and shims assembly (A) exploded from the valve housing.

69. Plug
70. Cap nut
71. Washers
72. Jam nut
73. Throttle valve screw
78. Adjusting washers
83. Metering screw

Fig. JD185—Checking operating pressure on series A, B and G "Powr-Trol" system. The shut-off valve must be located between the pressure gage and the valve housing on the return line.

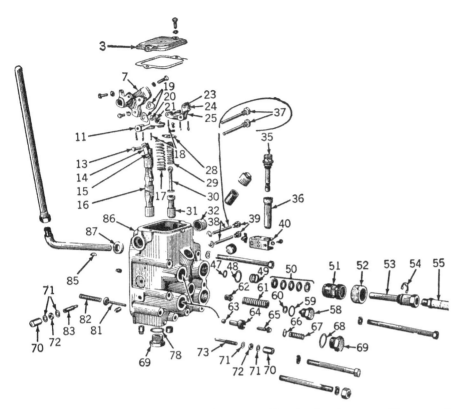

Fig. JD187—Exploded view of a typical "Powr-Trol" valve housing assembly as used on series A, B and G.

3. Cover	25. Cam latch arm	50. V-seal assembly	67. Spring
7. Cam	with roller	51. Adapter	68. Washer
11. Cam follower arm	28. Retainer	52. Adapter nut	69. Plug
with roller	29. Spring	53. Oil line coupling	70. Cap nut
13. Pin	30. Latch rod	54. Shear washer	71. Washers
14. Link	31. Release valve	55. Oil line	72. Jam nut
15. Rivet	32. Cap	58. Relief valve spring	73. Throttle valve
16. Operating valve	35. Oil line plug	plug	screw
with link	36. Cap	59. Gasket	78. Adjusting washers
17. Spring	37. Pivot pins	60. Adjusting washers	81. Throttle valve
18. Cotter pins	38. Pivot	61. Spring	82. Spring
19. Cam blade	39. Cap screw	62. Relief valve	83. Metering screw
20. Pin	40. Plug bracket	63. Ball	85. Woodruff key
21. Roller	47. Packing	64. Outer check valve	86. Valve housing
23. Cam latch pin	48. Washer	65. Inner check valve	87. Oil seal
24. Cam latch roller	49. Spring	66. Indexing washer	

from the valve housing, install the leak detector and close off the upper hose adapter opening. Apply air pressure to the valve stem (30 psi is sufficient) and observe the pressure gage to see if pressure is maintained. Check the lower check valve in the same manner except close off the lower hose adapter opening.

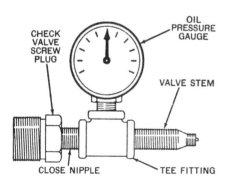

Fig. JD189—Home-made detector used for checking leaks in series A, B and G "Powr-Trol" valve housing check valves.

If the check valves will not hold air pressure, remove the valves and thoroughly clean them. If poppet type check valves and seats are slightly scored, they can be cleaned up by lapping with a very fine compound. If ball type check valves are leaking renew the balls and/or seats. Install check valves as outlined in the following paragraph.

359. INSTALL CHECK VALVES. Early production valve housings were factory equipped with poppet type check valves; whereas late and current production tractors are equipped with ball type check valves. On poppet type valves, Fig. JD191, install the inner

Fig. JD191—Series A, B and G early model check valve. The proper size hole in outer check valve (64) must align with the flat spot on the inner check valve (65). The proper register is obtained with indexing washer (66).

check valve with flat spot on outer end in register with the proper size hole in the outer check valve and secure in this position with the indexing washer (66). The small hole is used for series B; the medium size hole for series A; and the large hole for series G.

On the ball type check valves (Fig. JD193) as used on series A, B & G with 2½" or 3" remote cylinders, install inner check valve with short relief end toward outside which will permit one relief passage to open as shown in upper view. On series G with a 4" remote cylinder, install inner check valve with long relief toward outside which will permit both relief passages to open as shown in lower view.

360. ADJUSTMENT. To adjust either of the inner check valves, remove plug (69 — Fig. JD187) and valve lock spring. Insert valve lock (the lock is available from John Deere) in the check valve plug and reinstall the plug in the valve housing. Tighten the plug until both the inner and outer check valves are against their seats as shown in Fig. JD195B. At this time, there should be at least ¼ inch movement of the "Powr-Trol" lever when moving from neutral to raise position. If there is less than ¼ inch movement, grind off about 0.001 from inner end of inner check valve and recheck the lever movement.

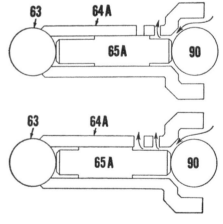

Fig. JD193—Series A, B, and G late model check valve. Correct installation of the inner check valve (65A) is explained in paragraph 359.

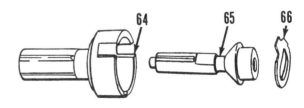

With the inner check valve adjusted, remove the check valve plug, valve lock and inner check valve. Reinstall the check valve plug and valve lock, leaving out the inner check valve. Tighten the plug until outer check valve is against its seat as shown in Fig. JD195C. Remove the valve housing cover and move the control lever until the cam blade comes against the cam follower roller (21—Fig. JD187). There should be ⅜-½ inch movement of control lever from this point until the cam on the operating valve tightens against steel ball and outer check valve. If movement is less than ⅜ inch, loosen the cam blade lock

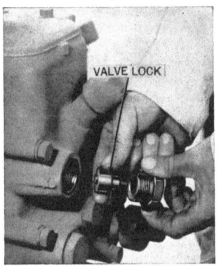

Fig. JD195A—When checking and adjusting John Deere "Powr-Trol" check valves, it is necessary to obtain the valve lock from John Deere.

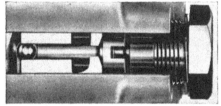

Fig. JD195B—Cut-away view of Deere "Powr-Trol" valve housing showing the valve lock installed for checking the inner check valve.

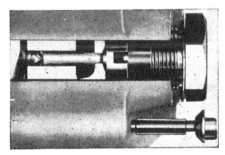

Fig. JD195C—Cut-away view of Deere "Powr-Trol" valve housing showing the valve lock installed for checking the outer check valve.

screw and move blade **in.** If lever movement is more than ½ inch, move the cam blade **out.** CAUTION: Hold blade firmly against milled slot when making the adjustment as in Fig. JD198.

Repeat the above procedure, for checking and adjusting the other check valve.

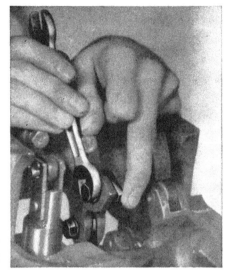

Fig. JD198—Adjusting cam follower blades on John Deere "Powr-Trol" valve housing.

OPERATING VALVE
Series A-B-G

361. **ADJUST.** To adjust the position of the operating valve, place control lever in neutral position and remove the valve housing cover. Place operating valve travel gage (the travel gage is available from John Deere) in the position shown in view (A—Fig. JD200), with shoulder of gage stem against top of operating valve. Move control lever to lowered position—at which time, the end of the gage stem should just slide over the end of the operating valve as shown in view B. If the opposite end of operating valve strikes plug (69—Fig. JD183) before end of travel gage stem slides over end of operating valve, add washers (78), which are available in thicknesses of 0.020 and 0.030. If the operating valve travels too far, remove washers to obtain correct travel.

Move control lever to neutral position and set the travel gage as shown in view (B—Fig. JD200) with end of gage just touching end of operating valve. Move control lever to raise position and gage stem to left—at which time the shoulder on gage stem should just slide over end of operating valve as shown in view (C). If operating valve has traveled too far, or not enough, adjust stop screw as shown in Fig. JD202 until desired travel is obtained.

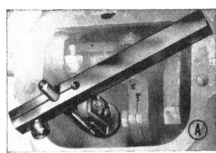

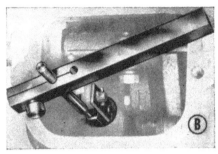

Fig. JD200—John Deere special travel gage positions for checking the "Powr-Trol" operating valve.

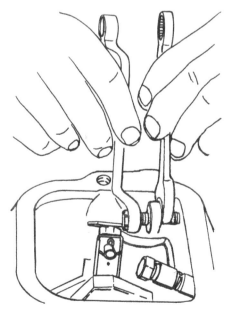

Fig. JD202—Adjusting John Deere "Powr-Trol" stop screw which controls the operating valve travel from neutral to raise position.

VALVE HOUSING

Series A-B-G

362. R&R AND OVERHAUL. To remove the "Powr-Trol" valve housing from tractor, drain oil from hydraulic system and remove valve housing from basic housing.

To disassemble the unit, remove cover, oil line adapters and pipe plugs from housing. Remove check valves. Remove the cam lock screw, bump control shaft out until Woodruff key centers in slot of cam forging, then drift out the Woodruff key as shown in Fig. JD204 and remove the control shaft. Compress the release valve spring and cam follower spring and lift out the cam and operating valve. Remove cam follower pivot pin and lever and cam latch pivot pin. Remove cam latch arm and spring. Remove the relief valve. The need and procedure for further disassembly is evident after an examination of the unit and reference of Fig. JD187.

Reassemble the valve housing by reversing the disassembly procedure and adjust the components as outlined in paragraphs 357, 360 and 361.

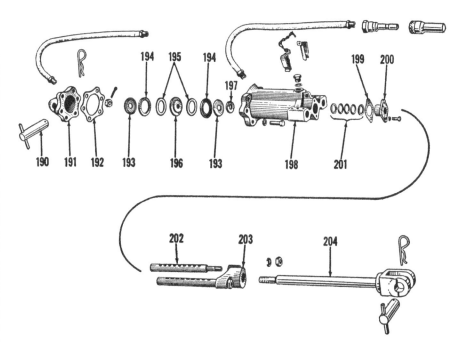

Fig. JD206—Exploded view of typical John Deere remote control cylinder. The working stroke is adjustable from 3-8 inches.

190. Attaching pin	195. Paper washer	200. Seal retainer
191. End cap	196. Piston	201. "V" seal assembly
192. Gasket	197. Spacer	202. Rod
193. Retainer	198. Cylinder	203. Piston rod stop
194. "U" cup packing	199. Shims	204. Piston rod and yoke

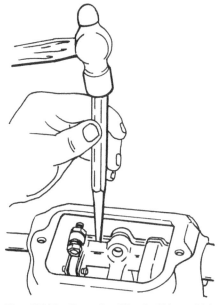

Fig. JD204—Removing Woodruff key which positions John Deere "Powr-Trol" control shaft in the cam forging.

REMOTE CONTROL CYLINDER

Series A-B-G

363. ADJUSTMENT. The double acting remote control cylinder is equipped with manual adjustments to vary the working stroke from 3 inches minimum to 8 inches maximum, in ¼ inch steps. CAUTION: Never attempt to operate the remote control cylinder unless both stop rod pins are in place.

364. DISASSEMBLE AND OVERHAUL. Remove oil lines and end cap (191—Fig. JD206). Remove the piston retaining nut and piston assembly, taking care not to damage the Neoprene piston rings. Withdraw piston rod and yoke from cylinder. Remove piston rod stop (203), adjusting rods (202), seal retainer (200) and "V" seal assembly (201).

Inspect face of end cap for nicks or burrs, adjusting rods for being bent and piston rod for burrs, scratches and/or being bent. Bent adjusting rods will be O.K., providing they can be thoroughly straightened. Small burrs and/or scratches can be removed from piston rod by using a fine hone; how-ever, piston rod should be renewed if it is bent. Renew any other questionable parts.

Lubricate the piston rod and reassemble the cylinder, leaving out the Neoprene piston rings. Attach a spring scale as shown in Fig. JD208 and check the pull required to move the lubricated piston rod through the "V" seal assembly. Add or deduct shims (199—Fig. JD206) until a pull of approximately 4 pounds is required. Shims are 0.010 and $\frac{1}{32}$-inch thick. When adjustment is as specified, install the Neoprene piston rings and tighten the piston retaining nut. Using a new gasket, install the end cap and tighten the retaining cap screws to a torque of 125 ft.-lbs.

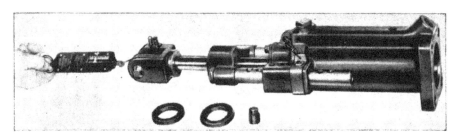

Fig. JD208—Checking pull required to move lubricated piston rod through "V"—seal assembly on Deere remote control cylinder.

Styled model D tractor. The front end support is bolted to the bottom of the cylinder block.

NOTES

SHOP MANUAL
JOHN DEERE
MODELS M-MT

IDENTIFICATION

Tractor serial number stamped on plate at base of instrument panel.

TRACTOR SERIAL NUMBERS BUILT SOMETIME IN THE FOLLOWING YEARS

Model	M	MT
1947	10,000	. . . . .
1948	20,000	. . . . .
1949	30,000	10,000
1950	40,000	20,000

Model M is produced in two versions: Adjustable and non-adjustable axle types.

Model MT is produced in three versions: Single and double wheel tricycle, and adjustable axle types.

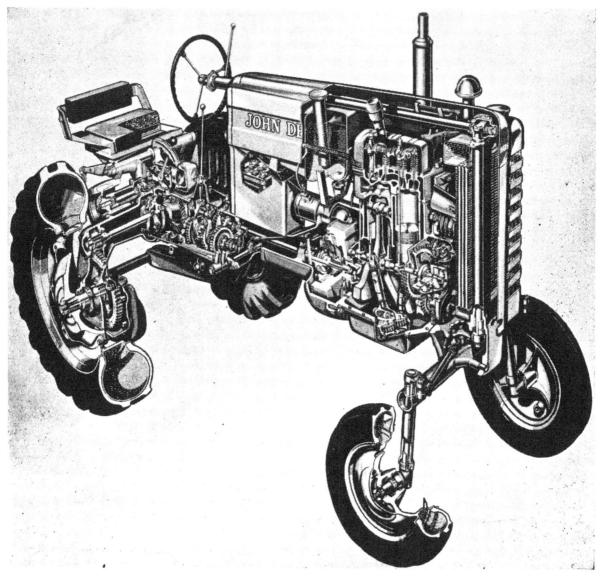

JOHN DEERE MODEL M

INDEX (By Paragraphs)

CONDENSED SERVICE DATA

GENERAL

Tractor Model	M	MT
Engine		
Make	Own	Own
No. Cylinders	2	2
Bore—Inches	4	4
Stroke—Inches	4	4
Displacement—Cubic Inches	101	101
Compression Ratio	6.0 to 1	6.0 to 1
Cylinders Sleeved?	No	No
Pistons Removed From? ...	Above	Above
Main Bearings, Number of..	2	2
Main Bearings Adjustable?..	No	No
Rod Bearings Adjustable?..	No	No

Tractor Model	M	MT
Speeds		
Number Forward	4	4
Number Reverse	1	1
Electrical System		
Voltage	6	6
Ground Polarity	Pos.	Pos.
Generator Make	D-R	D-R
Generator Model	1101857	1101857
Regulator Make	D-R	D-R
Regulator Model	1116816	1116816
Starting Motor Make.....	D-R	D-R
Starting Motor Model....	1107064	1107064

TUNE UP

Tappets & Valves	M	MT
Tappet Gap (Intake & Exhaust)......	0.012C	0.012C
Valve Seat Angle (Intake)...	30°	30°
Valve Seat Angle (Exhaust)...	45°	45°
Valve Stem Diameter......	0.372	0.372
Ignition & Spark Plugs		
Type	Bat.	Bat.
Distributor Make	D-R	D-R
Distributor Model	1111709	1111709
Breaker Gap	0.020	0.020
Static Timing	*	*

	M	MT
Plug Make	Champion	Champion
Plug Model (Gasoline)......	H10	H10
Electrode Gap	0.025	0.025
Carburetor		
Make	M-S	M-S
Model	TSX245	TSX245
Float Setting	¼	¼
Governed Speed		
Crankshaft—No Load rpm......	1825	1825
Belt Pulley—No Load rpm......	1746	1746
P.T.O.—No Load rpm......	609	609

SIZES—CAPACITIES—CLEARANCES

Crankshaft & Bearings	M	MT
End Play (Thousandths)......	3-7	3-7
Main Journal Diameter......	2.3975	2.3975
Rod Journal Diameter (Crankpin)......	2.2495	2.2495
Rod Bearing Clearance (Thousandths)......	2-4	2-4
Main Bearing Clearance (Thousandths)......	1-3.5	1-3.5
Tightening Torques (Ft.-Lbs.)		
Cylinder Head Cap Screws......	105	105
Rod Bolts......	55-60	55-60
Main Bearing Bolts......	140	140
Pistons-Pins-Rings		
Piston Skirt Clearance......	0.003	0.003
Piston Pin Diameter......	1.1878	1.1878
Ring End Gap (Thousandths)......	10-20	10-20

Camshaft & Bearings	M	MT
End Play (Thousandths)......	3-7	3-7
Journal Diameter	1.810	1.810
Bearing Clearance (Thousandths)......	1.5-3.5	1.5-3.5
Liquid Capacities		
Cooling System	3½ Gals.	3½ Gals.
Transmission & Differential	6½ Qts.	7½ Qts.
Crankcase	5 Qts.	5 Qts.
Fuel Tank	9.5 Gals.	9.5 Gals.
Touch-O-Matic	3 Qts.	5 Qts.
Final Drives—each	3½ Pts.	2 Qts.
Belt Pulley	½ Pt.	½ Pt.
Front Wheels		
Toe-in	1/8-3/16 In.	1/8-3/16 In.

*Timed with distributor cam in fully advanced position to mark "SPARK" on flywheel.

FRONT SYSTEM AND STEERING
(Axle and Tricycle Types)

Model M tractors are available in either adjustable or non-adjustable axle versions only; whereas, Model MT tractors, in addition to being available in the single and dual wheel tricycle types, are also available in an ajustable axle version.

LOWER SPINDLE OR FORK

Model MT

1. **R & R AND OVERHAUL.** Removal of either the fork and front wheel assembly (Fig. JD301) on the single front wheel version or of the lower spindle and knuckle and wheels assembly (Fig. JD300) on the dual wheel versions is accomplished by supporting front of tractor and removing cap screws (27) which retain the fork or lower spindle to the upper spindle. These special cap screws (27) are heat treated and have Allen heads. CAUTION: Do not use ordinary cap screws at this point.

27. Screws (heat treated)
35. Dowels
36. Wheel fork
37. Wheel axle
38. Outer bearing
39. Grease seal
40. Adjusting shims
41. Bearing retainer
42. Rim half
43. Axle clamp
44. Oil seal
45. Bearing retainer
46. Hub & rim

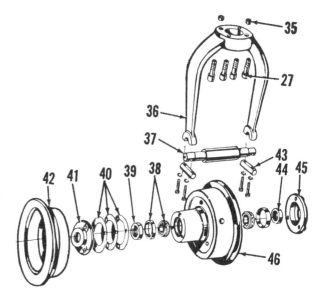

Fig. JD301—Lower assembly of front end unit for Deere MT tricycle type tractor having a single front wheel. Wheel fork (36) is bolted to upper spindle in the same manner as the lower spindle (26) assembly shown in figure JD300.

Front wheel bearings on the single wheel fork mounted version should be adjusted to 0.000-0.003 preload by varying the number of shims (40— Fig. JD301) which are located between bearing retainer (41) and hub (46).

STEERING KNUCKLES

Models M-MT

2. To remove knuckles from axle or axle extensions, support front of tractor and remove wheels and bearings assemblies. On early production M tractors, remove nuts from spindle lock pins (77—Fig. JD305) and drive pins out of steering arms (55). On late production M tractors and all MT tractors, remove cap screws (57-Figs. JD304 and 306) and bump steering arms from knuckles. Knuckles can then be withdrawn from below and needle bearings (69) can be driven out of axle or axle extensions.

New needle bearings should be pressed or driven in with a special driver such as John Deere tool number AM-457-T. Apply tool against end of bearing which has name stamped on it. When reassembling, make certain that tangs on thrust washer (64) fit through slots in seal retainer washer (66) and into notches in flange of knuckle. On late production tractors,

26. Lower spindle
27. Screws (heat-treated)
29. Seal
31. Inner bearing
32. Wheel hub
33. Outer bearing
34. Hub cap
35. Dowels

Fig. JD300—Lower spindle for Deere MT tricycle type tractor having dual front wheels. Item (26) is bolted to upper spindle by means of heat-treated screws (27) in the same manner as the tricycle fork (36) shown in figure JD301.

adjust up and down play of knuckles to 0.000-0.030 (I&T recommended) by varying the number of shims (54) which are available in thicknesses of 0.010 and 0.027.

TIE RODS & TOE-IN

Models M-MT

3. Recommended toe-in is ⅛-³⁄₁₆ inch on non-adjustable axles, ⅛-½ inch on adjustable axles. Adjustment of toe-in is made by loosening the tie rod clamps and rotating tie rods on tie rod ends. Tie rod ends are rubber mounted and non-adjustable for wear.

Note: Steering gear must be in its mid, or high point position and front wheels must be pointing straight ahead when making the toe-in adjustment.

AXLE PIVOT SHAFT & BUSHINGS

Models M-MT

4. Early production model M front axle pivot shaft was carried in two steel-backed, rubber lined bushings. When servicing these units, obtain and install a "Front Axle Pivot Shaft and Bushing Change-over Kit", which contains bronze bushings and a new style ¹³⁄₁₆-inch diameter pivot shaft.

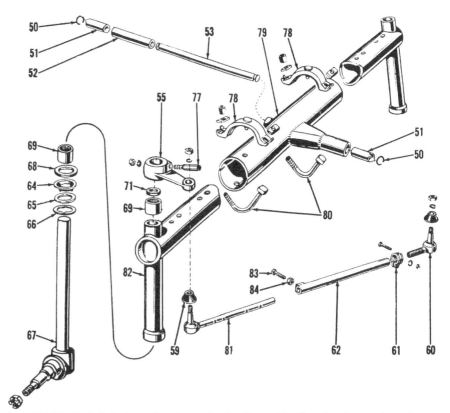

Fig. JD305—Exploded view of early production Deere M adjustable front axle and associated parts. Late production units are similarly constructed except that the steering arm is retained to the spindle by a cap screw. See Fig. JD304. Some models use heavy duty type clamps (78) as shown in Fig. JD306.

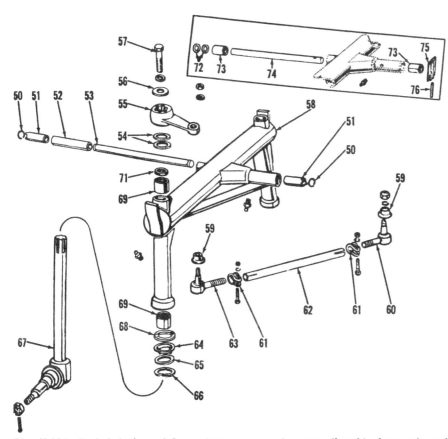

Fig. JD304—Exploded view of Deere M late production non-adjustable front axle and associated parts. The front axle pivot shaft and bushings change over kit is shown in the box. Early production units were similar except that the steering arm was retained to the spindle by a spindle lock pin. See Fig. JD305.

Late production M tractors and all MT tractors are factory equipped with a 1-inch diameter pivot shaft and bronze bushings.

5. To renew the front axle pivot shaft and/or bushings, remove axle main member from tractor and drive bushings out of axle main member. After installing bushings, ream them if necessary to provide 0.0025-0.005 (I&T recommended) clearance for the pivot shaft.

R&R COMPLETE FRONT UNIT

Models M-MT

6. To R&R the complete front system, including wheels, front support, spindle or fork or axle and steering gear as a single unit, remove hood, grille and air cleaner. Drain

50. Snap ring	66. Seal retaining washer
51. Rubber lined pivot bushing	67. Spindle and knuckle
52. Bushing spacer	68. Thrust washer
53. Pivot shaft	69. Needle bearing
54. Shims (0.010 & 0.027)	71. Felt seal
55. Steering arm	72. Thrust washers
56. Washer	73. Bushing
57. Cap screw	74. Pivot shaft
58. Axle	75. Lock plate
59. Dust cover	76. Groov pin
60. Tie rod end	77. Spindle lock pin
61. Clamp	78. Axle clamp
62. Tie rod tube	79. Axle center member
63. Tie rod end	80. U-bolt
64. Thrust washer	81. Adjustable tie rod end
65. Lower seal	82. Axle extension (knee)
	83. Set screw
	84. Jam nut

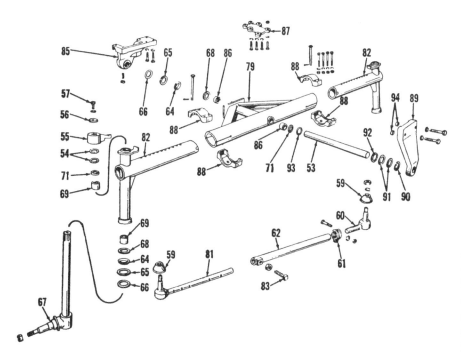

Fig. JD306—Exploded view of Deere MT adjustable front axle and associated parts.

53. Pivot shaft	65. Seal	85. Pivot shaft rear support
54. Shims	66. Seal retaining washer	86. Bushing
55. Steering arm	67. Spindle and knuckle	87. Steering lever
56. Washer	68. Thrust washer	88. Clamp
57. Cap screw	69. Needle bearing	89. Pivot shaft front support
59. Dust cover	71. Felt seal	90. Snap ring
60. Tie rod end	79. Axle center member	91. Spring washer
61. Clamp	81. Adjustable tie rod end	92. Washer
62. Tie rod tube	82. Axle extension	93. Seal retainer
64. Thrust washer	83. Set screw	94. Dowels

cooling system, disconnect radiator hoses and disconnect radiator from upper bracket. Remove Allen screws (1-Figs. JD311 and 313) and slide steering tube rearward until free of steering worm. Remove radiator, support front of tractor and remove cap screws retaining front support to engine and roll unit away from tractor.

R & R UPPER SPINDLE

Model MT

10. To remove the upper spindle (25-Fig. JD312), remove the gear unit as outlined in paragraph 14. Support front of tractor and detach wheel fork or lower spindle from upper spindle, or, on axle type tractors, detach axle from front support and center steering arm from upper spindle. The upper spindle can now be withdrawn from below after first removing the coupling sleeve pin (20-Fig. JD312). The upper spindle bushing (23) is presized and if carefully installed, will require no final sizing. Renew needle bearing (18) and/or thrust bearing (17) if corroded or worn.

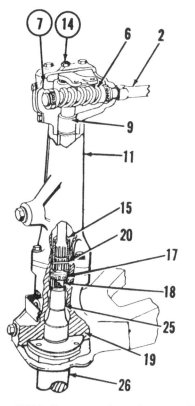

Fig. JD310—Upper portion of Deere MT front end unit. Item (26) can be either a lower spindle for dual wheels, a fork for a single front wheel or a tie rod plate for an adjustable front axle. Refer to legend under Figs. JD312 & 313.

Fig. JD311 — Exploded view of Deere M steering gear, front support and associated parts. Shims (7) control worm shaft end play.

1. Allen screw
2. Steering tube
3. Seal
4. Upper bushings
5. Bearing assembly
6. Worm
10. Vertical shaft upper seal
11. Shaft tube
12. Vertical shaft
13. Top cover
14. Adjusting screw
15. Gasket
16. Lower bushing
17. Vertical shaft lower seal
18. Center steering arm
19. Front support
21. Gear housing rear cover
22. Gear housing
23. Expansion plug

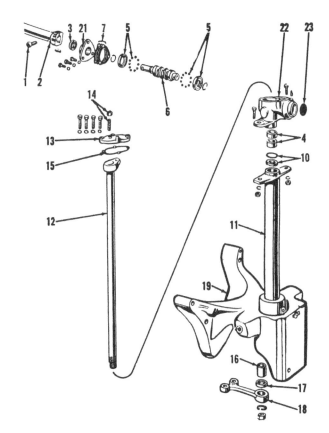

STEERING GEAR ASSEMBLY

For the purposes of this manual, the MT steering gear assembly (Fig. JD313) will include the gear housing and all parts contained therein. The M steering gear assembly (Fig. JD311) will include the shaft tube, gear housing and all parts contained therein. For methods of removal of the front support refer to paragraph 6.

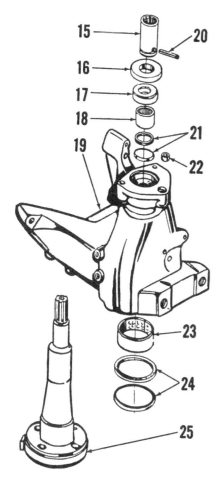

Fig. JD312—Exploded view of Deere MT front support.

15. Coupling
16. Thrust washer
17. Thrust bearing
18. Needle bearing
20. Groov pin
21. Seal
22. Dowel
23. Bushing
24. Seal
25. Upper spindle

Models M-MT

The cam and lever type gear is provided with two adjustments; one for worm-shaft end play, and the other for mesh or backlash. Refer to Figs. JD311 and 313.

12. ADJUST WORM SHAFT END PLAY. To remove any noticeable end play on Model M, vary the number of shims (7—Fig. JD311) between the steering gear housing rear cover and the housing; on Model MT, vary the number of shims (7—Fig. JD313) between the steering gear housing front cover and the housing. When properly adjusted, the shaft should have zero end play, yet should rotate freely or with only a perceptible drag when turned through the mid position.

13. ADJUST BACKLASH. With the wormshaft bearings adjusted as outlined in paragraph 12, support front of tractor and place the vertical (lever stud shaft) shaft on the high point of the worm by rotating the steering wheel to its mid-position, half way between full right turn and full left

3. Seal
4. Bushings
5. Bearing assembly
6. Worm
7. Shims
8. Front cover
10. Vertical shaft seal
11. Gear support
12. Vertical shaft
13. Top cover
14. Adjusting screw

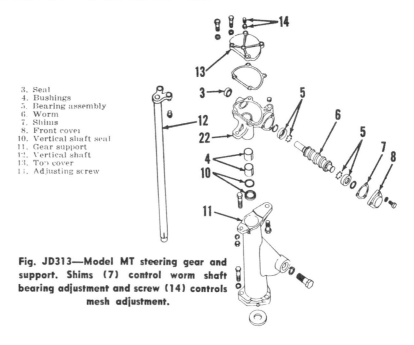

Fig. JD313—Model MT steering gear and support. Shims (7) control worm shaft bearing adjustment and screw (14) controls mesh adjustment.

turn. Now turn adjusting screw (14—Fig. JD311 and 313) in until a barely perceptible drag is experienced only when rotating steering wheel through the mid-position; it should turn freely and should have a slight amount of backlash at all other positions.

14. R&R AND OVERHAUL. To remove the steering gear unit, first remove hood, grille and Allen screw connecting the steering wheel tube to the worm shaft. Slide tube rearward until same is free from worm. On model M, disconnect center steering arm from vertical steering shaft and remove bolts retaining steering gear housing to the vertical shaft tube. On model MT, remove cap screws joining lower end of steering gear support (11—Fig. JD313) to front support. Lift gear unit from tractor.

Disassembly is evident after an examination of the unit and reference to Figs. JD311 and 313. Bushings (4) should be honed or reamed to 0.999-1.000 inside diameter after installation.

ENGINE AND COMPONENTS

R&R ENGINE WITH CLUTCH

20. To remove engine, as shown in Fig. JD320, remove hood and grille. Detach steering wheel tube from steering worm shaft. Disconnect fuel line, heat indicator, controls and wires from engine. On tractors equipped with Touch-O-Matic, disconnect oil lines from top of instrument panel and remove the lines. Support rear of tractor under center frame and engine separately. Remove flywheel dust cov-er. Remove cap screws, bolts and nuts attaching engine to center frame and slide engine forward and clear of propeller (clutch) shaft. With the engine supported in a hoist, remove radiator. Unbolt and remove front support and front axle assembly as a unit.

CYLINDER HEAD

21. The first step in removing the head is to remove the hood and disconnect throttle controls. Remove air inlet pipe, carburetor and manifolds.

Disconnect oil line and heat indicator from head. Remove tappet levers and shaft assembly.

Remove tappet rods (push rods). Remove cap screws retaining head to block and remove head.

Tighten head cap screws evenly and progressively from center outward and to a torque value of 105 foot pounds. If tappet lever shaft support studs were removed, be sure rear stud (3—Fig. JD321) is installed in the cor-

Fig. JD320—M and MT engine removal. Note support under center frame and balance point of engine hoist.

rect location. The rear drilled stud provides an oil passage for lubrication of tappet levers. Reinstall tappet lever shaft so that the oil holes face down and towards the head. Adjust tappet clearance to 0.012 cold.

VALVES AND VALVE SEATS

22. Tappets should be set cold to 0.012 for both inlet and exhaust valves. Inlet and exhaust valves, which seat directly in the cylinder head have a stem diameter of 0.3715-0.3725, and are not interchangeable. The inlet valve has a face and seat angle of 30 degrees with a desired seat width of ⅛-inch. The exhaust valve has a face and seat angle of 45 degrees with a desired seat width of 5/64-inch. Seats can be narrowed, using a 20 and 70 degree cutter or stone.

VALVE GUIDES

23. The (0.3745-0.376) inlet and exhaust valve service guides (8—Fig. JD321) are interchangeable. Guides should be pressed or driven into the cylinder head, using a piloted drift, until port end of guide is $2\frac{5}{32}$ inches from gasket surface of cylinder head. Ream guides after installation to obtain the desired stem to guide clearance of 0.002-0.0045.

VALVE SPRINGS

24. Inlet and exhaust valve springs are interchangeable. Each spring should require 39-45 pounds pressure

to compress it to a height of 2 inches. Springs which are rusted, distorted, or do not meet the foregoing pressure specifications should be renewed.

VALVE TAPPET LEVERS (ROCKER ARMS)

25. To remove valve tappet levers and shaft assembly, remove hood and tappet cover (rocker arm cover). Remove nuts retaining tappet lever shaft supports to cylinder head and lift assembly from head. To disassemble, pull one of the end caps off shaft, then engage spring (11—Fig. JD321) with a wire hook and lift off the end cap (5) as in Fig. JD322. Tappet levers, supports and springs can then be removed from shaft.

Valve contacting surface of tappet lever can be refaced but original arc must be maintained by performing this operation on a fixture especially designed for refacing tappet levers. Diameter of tappet lever shaft is 0.810-0.8105. Tappet lever bore (inside diameter) is 0.812-0.813. Clearance in excess of 0.006 between unbushed bore of lever and shaft is corrected by renewing tappet lever and/or shaft.

Inlet and exhaust tappet levers are not interchangeable. When reassembling, reinstall tappet lever shaft so that oil holes face down toward the cylinder head. If shaft support studs have been removed, be sure they are reinstalled in their correct location,

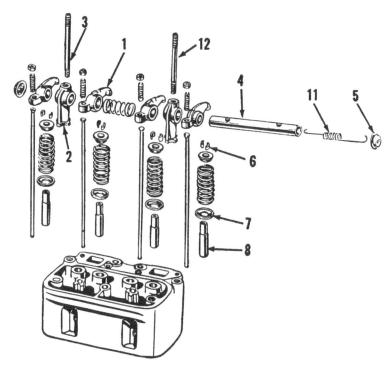

Fig. JD321—M and MT Deere tappet levers and shaft assembly. Rear stud (3) provides oil passage for tappet lever lubrication.

1. Tappet lever	5. End cap
2. Shaft support	6. Split cone lock
3. Oil passage stud	8. Valve guide
4. Tappet lever shaft	11. Retaining spring

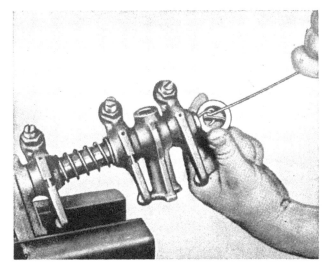

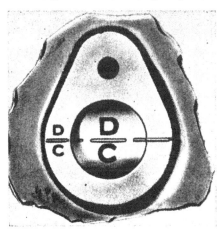

Fig. JD322—Using hooked tool to disassemble M and MT tappet levers assembly. Tool can be made from welding rod.

rear stud (3—Fig. JD321) is drilled to provide an oil passage for lubrication of tappet levers.

CAM FOLLOWERS

26. The mushroom type cam followers operate directly in machined bores of the cylinder block. Cam followers are supplied only in the standard size and should have a suggested clearance of 0.0005-0.0025 in block bores. Any tappet can be removed after removing the camshaft as outlined in paragraph 30.

VALVE TIMING

27. Inlet valve opens at top dead center and exhaust valve closes 5 degrees after top dead center. Valves are correctly timed when mark on camshaft gear is meshed with an identical mark on crankshaft gear as in Fig. JD325.

To check valve timing when engine is assembled, adjust tappets to 0.012 clearance. Turn engine until front cylinder inlet valve is just opening. At this time the D/C mark on flywheel, shown in Fig. JD326, should be within ¼ inch of index line on side

of timing hole. If marks do not register within ¼ inch, remove camshaft and remesh camshaft gear with crankshaft gear.

TIMING GEAR COVER

28. To remove timing gear cover, remove radiator, front support and front axle as in paragraph 6. Remove fan and bracket assembly. Remove generator and, if so equipped, hydraulic pump unit. Disconnect controls and remove governor. Remove pulley from

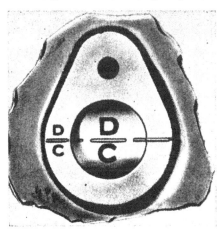

Fig. JD326—M and MT top center position is when flywheel mark DC registers with index on right side of engine.

crankshaft and cover to cylinder block cap screws. When re-installing cover, refer to note in paragraph 29.

CRANKSHAFT FRONT OIL SEAL

29. To renew crankshaft front oil seal. remove radiator, front support and front axle as per paragraph 6. Remove fan and bracket assembly. Remove pulley from crankshaft. Remove front oil seal plate (14—Fig. JD330). Oil seal (15) can be pressed out of plate and renewed.

Note: Soak new seal in oil and install in plate with lip of seal facing toward rear of tractor. Slide seal and plate assembly on crankshaft using a piece of feeler stock as a sleeve to prevent damaging seal.

CAMSHAFT AND BEARINGS

30. To remove camshaft, first remove the timing gear cover and tappet levers (rocker arms) assembly. Remove oil pan, oil pump and ignition distributor. Unlock thrust plate cap screws shown in Fig. JD329. Remove screws from block. Hold cam followers up and carefully withdraw camshaft. To remove gear from shaft,

Fig. JD329—Removing M and MT cap screws which retain camshaft thrust plate to cylinder block. Cap screws have tab type lock washers.

Fig. JD325—M and MT valve timing marks on camshaft and crankshaft gear.

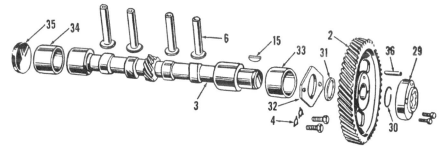

Fig. JD328—M and MT camshaft, bushings and cam followers. Hydraulic pump drive coupling (29) is bolted to camshaft gear.

2. Camshaft gear	6. Cam follower	31. Spacer washer	34. Rear bushing
3. Camshaft	15. Gear key	32. Thrust plate	35. Expansion plug
4. Tab washers	30. Snap ring	33. Front bushing	36. Dowel pin

remove hydraulic pump drive (29-Fig. JD328) and snap ring (30); then press shaft out of gear.

Camshaft journals rotate in two renewable presized bushings. Shaft journal size is 1.8095-1.8105. Shaft clearance in bushings is 0.0015-0.0035. Inside diameter of bushings is 1.812-1.813. The maximum permissible clearance between camshaft journals in their bushings is 0.005 and when it exceeds this amount, it will be necessary to renew the bushings.

To renew camshaft bushings, remove engine and flywheel to gain access to expansion plug (35) behind rear bushing. Bushings can be driven out or pressed out of block and new ones pressed or driven in. Bushings are presized and, if carefully installed, using John Deere driver tool number AM-458-T or equivalent, require no final sizing. Be sure oil hole in each bushing registers with oil hole in cylinder block.

Position expansion plug (35) with cupped side of plug facing rearward. Press the plug into the block until front surface of plug is 13 7/32 inches, plus or minus 1/64, from front face of block at front camshaft bearing bore. The expansion plug must be installed in this position to assure an adequate oil supply to the tappet lever (rocker arm) oil passage.

Install thrust plate spacer washer (31) with chamfered edge of spacer bore facing front camshaft journal. Install thrust plate (32) and Woodruff key (15) and press gear on shaft until gear seats firmly against spacer washer. Length of spacer washer (31) and thickness of thrust plate control camshaft end play. For correct end play, the measured thickness of washer (31) should be 0.003 to 0.010 greater than the thickness of thrust plate (32).

Mesh single mark on camshaft gear with single mark on crankshaft gear as shown in Fig. JD325. Install oil pump as outlined in paragraph 39 and ignition distributor as in paragraph 51.

TIMING GEARS

31. **CAMSHAFT GEAR.** Gear is keyed to shaft and the press fit is usually such as to require removal of

the camshaft if the gear is to be renewed. The quickest way to remove the camshaft is to R&R engine from tractor.

CRANKSHAFT GEAR. To R & R the crankshaft gear it is first necessary to remove the crankshaft which in turn necessitates the removal of the engine from the tractor. Refer to paragraph 20 for engine removal. To facilitate installation of gear on crankshaft, heat gear in boiling water.

GOVERNOR GEAR. Removal requires governor removal; refer to GOVERNOR section.

ROD AND PISTON UNITS

32. Piston and connecting rod assemblies are removed from above after removing cylinder head and oil pan.

Remove carbon accumulation and ridge from unworn portion of cylinders to prevent damaging ring lands when pistons are withdrawn. Connecting rods are not numbered and should be identified in some manner for reassembly. Install connecting rods and caps so that raised marks face camshaft side of engine and tighten the connecting rod cap screws to 55-60 ft. lbs. torque.

PISTONS AND RINGS

33. Cast iron pistons are supplied in standard, 0.020 and 0.040 oversize. Diameter of the standard piston is 3.9965-3.9975 inches. Diameter of the standard cylinder bore is 4.000-4.001. Rebore cylinder if out-of-round more than 0.003. Clearance of piston skirt in cylinder should be 0.003. Raised mark on piston heads should face toward front of engine.

There are three compression rings and one ventilated oil ring per piston. All rings are stamped either "TOP" or with a small pip mark to indicate the top portion of the ring. Recommended end gap for all rings is 0.010-0.020. Recommended side clearance for all compression rings is 0.003; for the oil control ring, 0.001-0.0025. Rings are supplied in the standard, 0.020 and 0.040 oversize.

PISTON PINS

34. The 1.1877-1.1879 inch diameter floating piston pins which are retained in the piston bosses by snap rings, are available only in the standard size. Fit pins to a thumb press fit in piston and with a clearance of 0.0004-0.0009 in connecting rod bushing. Bushing inside diameter is 1.1883-1.1886.

CONNECTING RODS AND BEARINGS

35. Connecting rod bearings are steel backed, babbitt lined, non-adjustable, precision type and can be renewed without removing connecting rod and piston assembly. Replacement rods are not marked and should be installed with the small raised mark on rod and cap facing the camshaft. Bearings are available in 0.003 and 0.005 undersizes as well as standard.

Journal size 2.249-2.250
Running clearance 0.002-0.004
Cap screw torque........55-60-ft. lbs.

CRANKSHAFT AND BEARINGS

36. Crankshaft is supported on two steel backed, babbitt lined, non-adjustable, precision type bearings, renewable from below after removing the oil pan. Renew worn bearings; do not file caps. Remove main bearing caps and turn bearing upper halves out of cylinder block with a reworked rivet inserted in oil hole of crankshaft journal. Rear bearing controls crankshaft end play of 0.003-0.007. Bearings are available in the 0.003 and 0.005 undersizes as well as standard.

The quickest method of removing the crankshaft is to remove the engine from the tractor as outlined in paragraph 20.

Check crankshaft journals for wear, scoring and out-of-round condition against the values listed below:

Journal diameter 2.397-2.398
Running clearance 0.001-0.0035
Renew shaft if
out-of-round more than.......0.003
Cap screw torque........140 ft. lbs.

CRANKSHAFT REAR OIL SEAL

37. To renew the treated leather oil seal (3-Fig. JD330), which is mounted in a one-piece retainer (1), split tractor (detach engine from center frame), as outlined in paragraph 62 and remove flywheel. Install new oil seal with the lip of same facing towards front of engine. Soak seal assembly in oil before installing. Slide seal and

OIL PUMP

39. **REMOVE & REINSTALL.** To remove oil pump, remove flywheel dust shield and oil pan. Remove lock nut from oil pump attaching screw located on right side of upper crankcase, then the Allen screw, as shown in Fig. JD332 and withdraw pump from bottom of crankcase.

When reinstalling pump, rotate engine crankshaft until number one (front) cylinder is on compression stroke and the mark "DC" on flywheel indexes with mark at timing hole. Install pump and mesh gears (pump to camshaft) so that distributor driving groove in pump gear is parallel to the crankshaft and narrow part of groove wall is towards outside of tractor. Recheck ignition timing.

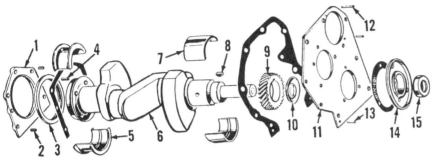

Fig. JD330—M and MT crankshaft and main bearings. Renewal of rear oil seal (3) requires detaching engine from center frame and removal of flywheel.

1. Oil seal retainer
2. & 4. Dowel pin
3. Oil seal
5. Rear bearing
6. Crankshaft
7. Front bearing
8. Gear key
9. Crankshaft gear
10. Oil slinger
11. Gear cover
12. Dowel pin (upper)
13. Dowel pin (lower)
14. Oil seal plate
15. Oil seal

Fig. JD332—M and MT oil pump is retained in position by an Allen head screw located on right side of upper crankcase.

plate assembly on crankshaft using a piece of feeler stock as a sleeve to prevent damaging seal.

Seal contacting surface of crankshaft must be smooth and true. Surface of crankshaft and bore of seal plate must be concentric within 0.005.

FLYWHEEL

38. To remove flywheel, split tractor (detach engine from center frame) and remove clutch as outlined in paragraph 62. Remove cap screws retaining flywheel to crankshaft and pry flywheel off dowels.

To install a new ring gear, heat same to 500 degrees F. and install gear with beveled edge of teeth facing rear of tractor.

There are two marks located on flywheel outer surface: "SPARK", as shown in Fig. JD340B, which is the ignition advanced running timing mark; and "DC", as shown in Fig. JD326, which is the top dead center mark of number one cylinder.

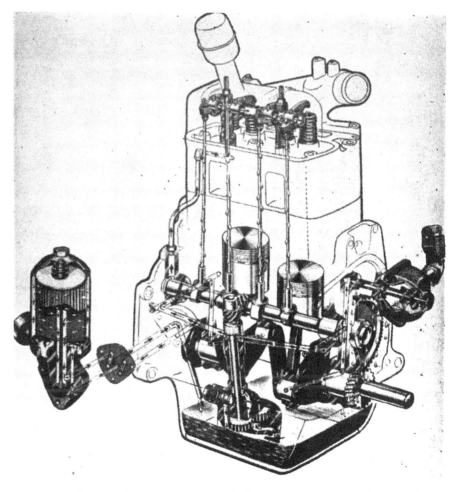

Fig. JD331—Phantom view of M and MT lubrication system. Rear tappet lever shaft support stud provides an oil passage to lubricate tappet levers assembly. Pressure relief valve (shim adjustment) is located in pump body.

40. OVERHAUL. End clearance between pump body gears and lower cover (11—Fig. JD333) should be 0.002-0.006. The recommended diametral clearance between the gears and pump body is 0.001-0.004. Check pump drive shaft and its mating surface for wear. New shaft diameter is 0.4825-0.4830 with a pump body bore diameter of 0.4835-0.4845. No gasket is used between lower cover and pump body; be sure surfaces are clean, true and not distorted.

OIL RELIEF VALVE

41. The spool type relief valve (3-Fig. JD333) is located in oil pump body and can be adjusted with shims (5) inserted between the spring and relief valve retaining plug to maintain a pressure of 30 psi. Oil pressure adjustment requires removal of the oil pan. The relief valve should fit in its bore with 0.003-0.005 of clearance.

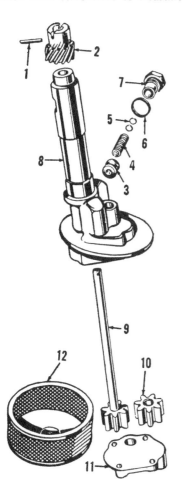

Fig. JD333—M and MT oil pump. Pressure relief valve adjustment is controlled by shim washers (5) at relief valve spring.

1. Gear pin	7. Plug
2. Drive gear	8. Pump body
3. Relief valve	9. Shaft & pump gear
4. Valve spring	10. Idler gear
5. Shims	11. Pump cover
6. Gasket	12. Pump screen

CARBURETOR

42. The carburetor is a Marvel-Schebler, TSX245. Carburetor must be removed to adjust the fuel level. Float should be set to ¼ inch when measured from bowl cover gasket surface to nearest edge of float. The idle mixture screw controls the flow of air in the idle system and turning the screw towards its seat richens the mixture. Approximate setting is ⅞ turn open. Power range is controlled by a power jet needle which reduces fuel flow and leans the mixture when turned towards its seat. Refer to Appendix 1 of this manual for overhaul data.

GOVERNOR

43. SPEED ADJUSTMENT. Adjust governor stop screw (SA-Fig. JD335) located in engine fan bracket to limit engine rpm as specified below:

Crankshaft rpm (Load).........1650
      (No Load) 1825
Belt Pulley rpm (Load)..........1580
      (No Load) 1746
Power Take-Off rpm (Load)..... 550
      (No Load) 609

44. LINKAGE ADJUSTMENT. Free-up and align all linkage to remove any binding. Adjust or renew any parts causing lost motion. Back out stop screw (SA-Fig. JD335) located in engine fan bracket. Place hand control

in full speed position and with carburetor throttle valve in wide open position adjust length of carburetor throttle rod so that the rod just enters hole in carburetor throttle shaft lever. Then shorten rod two turns. Disconnect hand lever control shaft to governor lever rod. With the hand control in the lowest speed position, push down on governor lever (18-Fig. JD-334). Now adjust rod connecting hand lever control shaft to governor lever

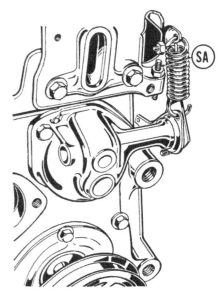

Fig. JD335—M and MT engine governed speed is controlled by adjusting screw (SA) located in left side of fan bracket.

1. Dowel pin (long)	11. Plug (right)
2. Dowel pin (short)	12. Front bushing
3. Rear bushing (in block)	13. Plug (lower)
4. Thrust washer	14. Plug (upper)
5. Shaft & gear	15. Housing
6. Weight pin	16. Seal retainer
7. Weight	17. Seal
8. Thrust bearing	18. Lever
9. Fork	19. Operating lever & shaft
10. Fork pin	20. Spring

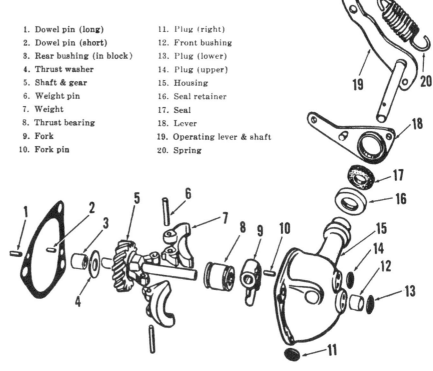

Fig. JD334—M and MT governor, exploded view.

so that holes in rod yoke and hole in governor lever are in register; then lengthen rod two turns. Adjust engine speed with governor stop screw as outlined in preceding paragraph.

45. R & R GOVERNOR. Remove hood and grille. Remove air cleaner and disconnect governor linkage. Remove fan and belts. Remove cap screws retaining governor housing to timing gear cover and remove housing. Withdraw governor shaft assembly being careful not to drop thrust washer (4-Fig. JD334) into timing gear case.

46. OVERHAUL GOVERNOR. Remove spring. Remove upper expansion plug (14). Bump pin (10) out of fork (9) and operating shaft. Remove expansion plug (11) from side of case and bump operating shaft out of case. Remove metal retainer (16) from side of case and renew oil seal. Weight pins (6) can be removed from carrier after grinding staked portion at end of pin. Shaft front bushing (12) can be renewed after removing lower front expansion plug (13). To remove shaft rear bushing (3) from cylinder block, tap bushing for ½ inch SAE thread and pull bushing with a cap screw. Bushings are presized and should be installed using a piloted drift.

After housing has been assembled, check condition of fork (9) and lever (19) by placing a new thrust bearing (8) under governor fork and laying case on a flat surface as shown in Fig.

JD335B. With operating lever moved upward (fork against bearing), center of hole in operating lever should measure 13/16 inch above flat surface. If this measurement is not within 1/64 inch, renew shaft and lever and/or fork. Reassemble and reinstall governor, and adjust engine speed as outlined in preceding paragraphs.

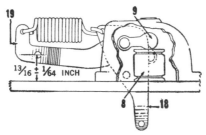

Fig. JD335B—Checking condition of Deere MT governor operating lever & shaft (19) and fork (9) using a flat surface and a new thrust bearing (8). Renew the lever and/or shaft if 13/16 inch measurement is not obtained.

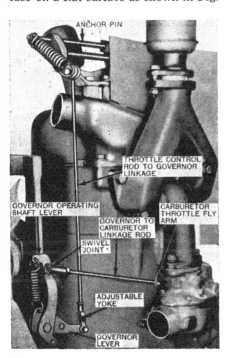

Fig. JD335A—M and MT governor and carburetor linkage, showing control rods and linkage adjustments.

COOLING SYSTEM
FAN AND RADIATOR

48. To remove fan assembly, remove hood and grille. Remove fan and generator belts and detach fan shaft from bracket.

Fan hub is not fitted with bushings. Correction of bearing wear is accomplished by renewing hub, bearing and/or shaft. See Fig. JD340.

49. Radiator can be removed without removing the front support assembly by the following procedure: Remove hood and grille. Remove Allen screw from steering tube and detach tube from steering gear worm shaft. Remove bolts from radiator upper bracket. Remove cap screws attaching radiator bracket to fan bracket. Detach lower right side bracket from radiator and lower left bracket from front support. Working on left side of tractor, carefully remove the radiator.

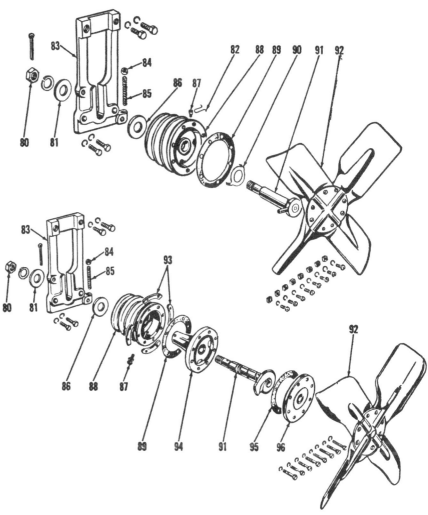

Fig. JD340—Exploded view of Deere fan and mounting bracket assemblies. Top view shows units as used on early production M tractors. Bottom view shows units as used on late production M tractors and all MT tractors.

80. Jam nut
81. Washer
82. Filler plug lock wire
83. Fan bracket
84. Governor stop screw jam nut
85. Governor stop set screw

86. Washer
87. Oil filler plug
88. Fan hub
89. Gasket
90. Thrust washer

91. Spindle
92. Fan
93. Double nut
94. Fan bearing
95. Gasket
96. Fan blade spacer

IGNITION & ELECTRICAL SYSTEM

50. John Deere M and MT are equipped with a model 1111709 Delco-Remy distributor and uses the Champion H10 spark plug with an electrode gap setting of 0.025.

IGNITION TIMING

51. Distributor rotation is clockwise when viewed from the drive end. Set breaker contacts to a 0.020 gap (32 deg. cam angle). To time the ignition, crank engine until No. 1 piston (front) is on compression stroke and the flywheel mark "SPARK" (advanced timing mark) aligns with timing port index as shown in Fig. JD340B. At this time, the distributor drive groove in oil pump drive gear should be parallel to crankshaft and with narrow part of groove wall towards outside of tractor.

Install distributor with retaining cap screws only finger tight. Rotate distributor cam as far as it will go in direction of normal rotation. While holding cam in fully advanced position, rotate distributor body in the opposite direction until points just begin to open. Tighten distributor retaining cap screws and install number one (front) cylinder spark plug wire in the terminal over the rotor contact.

As indicated above, the distributor is timed with the breaker cam in the

Fig. JD340B—M and MT flywheel mark "SPARK" is the fully advanced running timing mark of number one cylinder. Refer to Ignition Timing for correct timing procedure.

fully advanced position. To check action of the distributor governor, operate the engine at 600 rpm or less at which time the spark should occur at top center or when the DC mark on flywheel aligns with inspection port index line.

GENERATOR AND REGULATOR

52. Tractor is equipped with a Delco-Remy third brush generator model 1101857 and cut-out relay model 1116816. Generator output is controlled

by the third brush and a combination ignition and light switch. Switch has five positions: (1) Lights and ignition off; (2) Ignition on and low generator charging rate. In this position, a resistance is inserted in generator field circuit to reduce generator output to approximately two amperes; (3) Ignition on and high generator charging rate. This position removes resistance from generator field circuit and permits maximum generator output of approximately eight to ten amperes for charging a low battery; (4) Ignition and lights on and high generator charging rate; (5) Lights on, ignition off.

To adjust charging rate (8-10 amperes), set ignition-light switch at position number three (ignition on and high generator charging rate) and loosen third brush clamp screw on generator end frame; move third brush up to increase charging rate. Refer to Appendix 1 of this manual for overhaul data.

STARTING MOTOR

53. Starting motor is Delco-Remy model 1107064, with a Bendix drive. Starting motor is mounted on tractor center frame and its pinion engages flywheel ring gear from the rear.

CLUTCH AND CENTER FRAME

The structural element of the tractor located between the engine and the transmission is designated by the John Deere Co., as the center frame. This center frame functions as the clutch housing and to some extent as a torque tube. In subsequent paragraphs, it will be designated as the center frame with the other terms "torque tube" and "clutch housing" shown sometimes in parenthesis.

CLUTCH UNIT

The clutch is an Auburn 10-inch, spring loaded, single dry plate type, fitted with a 10001-1 cover assembly.

61. **ADJUSTMENT.** Clutch pedal should have 1½ inches free travel. Adjust free travel by loosening lock nut and turning adjusting cam (16-Fig. JD343) on front lower left side of center frame. Turning cam clockwise increases pedal free travel.

Fig. JD341—Deere M and MT clutch is a single plate 10 inch Auburn fitted with a 10001-1 cover assembly.

13. Actuating rod
16. Adjusting cam (pedal travel)
18. Pressure plate
19. Release bearing sleeve
20. Release yoke
21. Clutch shaft
23. Release lever
24. Lever adjustment screw
25. Release bearing
26. Pilot bushing
27. Lined plate
28. Flywheel
29. Center frame

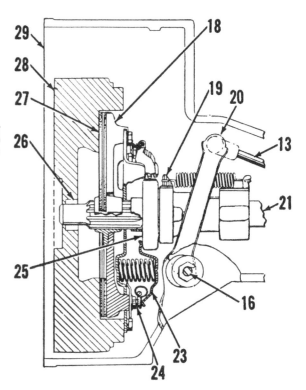

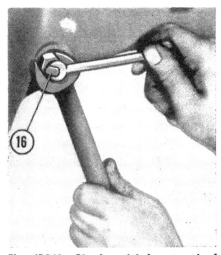

Fig. JD343—Clutch pedal free travel of 1½ inches is adjusted with a cam (16) located on left side of center frame. Rotating cam clockwise increases pedal free travel.

62. SPLITTING TRACTOR. To detach engine from center frame, as shown in Fig. JD320, proceed as follows: Remove hood and grille. Detach steering wheel tube from steering worm shaft. Disconnect fuel lines, heat indicator, controls and wires from engine. On tractors equipped with Touch-O-Matic, disconnect oil lines at front of instrument panel. Support rear portion of tractor under center frame and place engine in a hoist. Remove flywheel dust cover. Remove cap screws, bolts and nuts attaching engine to center frame and slide engine forward and clear of propeller (clutch) shaft.

62B. R&R CLUTCH. Split tractor as outlined in preceeding paragraph. Remove cap screws attaching clutch assembly to flywheel and remove cover assembly and driven plate.

Before reinstalling the clutch, pack the cavity in front face of flywheel one-half full of high temperature fibrous grease.

The long hub of lined plate should be installed away from flywheel. Align driven plate with a clutch aligning tool or a spare propeller (clutch) shaft when reinstalling clutch unit.

63. OVERHAUL CLUTCH. Refer to STANDARD UNITS section for complete clutch overhaul data.

Early production tractor flywheels were equipped with a bushing as the pilot bearing for the forward end of the clutch shaft; on late production tractors, the pilot bearing is of the shielded ball type. The early production bushing type can be converted to the late production ball type by obtaining and installing the late model clutch shaft, pilot bearing, grease retainer and pilot bearing adapter. Early model pilot bushing inside diameter is 0.991-0.992 with a clutch shaft pilot journal diameter of 0.987-0.988. Recommended clearance between bushing and clutch shaft is 0.003-0.005 with a maximum clearance of 0.010 for service.

64. OVERHAUL CLUTCH CONTROLS. To remove clutch pedal shaft, disconnect transmission from center frame as outlined in paragraph 71.

64A. Remove snap ring (1—Fig. JD-345) from right end of brake pedal shaft and remove right brake pedal (3). Bump pin (6) out of left brake pedal and remove pedal. Disconnect clutch control rod (13) from clutch control shaft (8) and withdraw brake pedal shaft (12). Remove bolt from clutch pedal and remove pedal. Clutch control shaft and spring can be withdrawn from inside of center frame. Bushings in center frame (torque tube) can be renewed at this time and, if carefully installed using a piloted drift, require no final sizing. Clutch pedal shaft bushing (10) inside diameter is 1.1211-1.1239. Brake shaft and pedal bushings (2 and 7) inside diameter is 0.8766-0.8794.

CENTER FRAME
(CLUTCH HOUSING)

65. R&R UNIT. To remove the center frame (clutch housing or torque tube) from the tractor, proceed as follows: Remove hood and grille. Detach steering wheel tube from steering worm shaft. Disconnect fuel lines, heat indicator, controls and wires from engine. On tractors equipped with Touch-O-Matic, disconnect rockshaft unit to instrument panel oil lines. Support rear portion of tractor under center frame and place engine in a hoist. Remove flywheel dust cover. Remove cap screws, bolts and nuts attaching engine to center frame and slide engine forward and clear of propeller (clutch) shaft. This completes the front split.

Fig. JD345 — M and MT center frame (torque tube) and components.

1. Snap ring
2. Brake pedal bushings
3. Brake pedal (right)
4. Brake pedal rod
5. Brake pedal (left)
6. Retaining pin
7. Pedal shaft bushing
8. Clutch pedal shaft
9. Clutch pedal spring
10. Bushing
11. Clutch pedal
12. Brake pedal shaft
13. Clutch actuating rod
14. Center frame
15. Timing hole cover
16. Starting motor retaining screw
17. Flywheel dust cover
21. Propeller (clutch) shaft
22. Groov pin

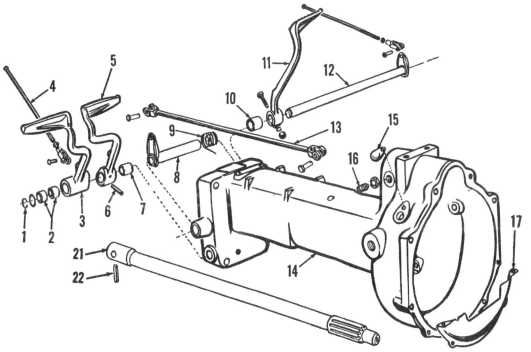

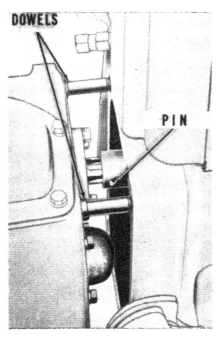

Fig. JD346—Separating transmission housing from center frame (clutch housing) to obtain access to input and output shaft bearing quills on models M and MT.

Disconnect brake rods. On tractors equipped with Touch-O-Matic, remove oil lines from rockshaft. Support both halves of tractor (transmission and center frame) separately and remove nuts and cap screws retaining transmission case to center frame. Slide front half of tractor forward (not more than two inches) on studs. Bump Groov pin out of propeller (clutch) shaft and transmission input shaft and move center frame away from transmission as shown in Fig. JD346.

66. OVERHAUL OR RENEW. To overhaul the clutch and brake pedal shafts proceed as outlined in paragraph 64A. The propeller (clutch) shaft (21-Fig. JD345) can be serviced at this time. If center frame is to be renewed, the procedure for transfer of miscellaneous parts from the old to the new unit is self-evident from a view of Fig. JD345.

TRANSMISSION

LOCAL REPAIRS

70. BASIC PROCEDURE. The transmission shafts, main drive bevel gears and differential unit are carried in the transmission housing. Although most transmission repair jobs involve overhaul of the complete unit, there are infrequent instances where adjustment, or the failed or worn part is so located that repair work can be com-

pleted safely without complete disassembly of the transmission. In effecting such localized repairs, time will be saved by observing the following as a general guide:

SHIFTER RAILS AND FORKS. Shifter rails and forks which are contained in the transmission top cover are accessible for overhaul after removing the cover.

OUTPUT SHAFT AND BEARINGS. The bearings for this shaft (23—Fig. JD350) can be adjusted after transmission is disconnected from the center frame (clutch housing) as outlined in paragraph 71. Adjustment is accomplished by removing the front bearing cone and adding or removing shims behind the bearing to obtain 0.005 preload. To remove the output

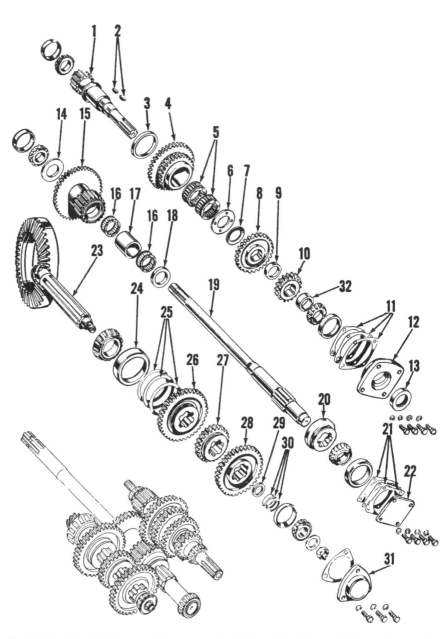

Fig. JD350—M and MT transmission assembly. Shims (11), (21), and (30) are for bearing adjustment. Shims (25) are for main drive bevel gears mesh adjustment.

1. Input shaft	13. Oil seal	24. Bearing cup
2. Gear key	14. Spacer washer	25. Shims
3. & 6. Spacer washer	15. Cluster gear	26. 2nd & reverse
4. 1st cluster gear	16. Gear bearing	sliding gear
5. Cluster gear bearing	17. Bearing spacer	27. 1st & 4th sliding
7. Spring washer	18. Spacer washer	gear
8. 4th drive gear	19. PTO shaft	28. 3rd sliding gear
9. Spacer	20. PTO coupling	29. Bearing spacer
10. 3rd drive gear	21. Shims	30. Shims
11. Shims	22. PTO shaft cover	31. Shaft cover
12. Bearing quill	33. Output shaft	32. Spacer

shaft, it is necessary to first disconnect transmission from center frame as per paragraph 71 and to remove both final drive assemblies and differential.

INPUT SHAFT AND BEARINGS. Bearings for this shaft (1) can be adjusted after detaching the transmission from the center frame as in paragraph 71. Adjustment is accomplished by removing or adding shims (11) under the bearing quill to obtain end play of 0.002-0.004. On M tractors before Ser. No. M-45073 and MT tractors before Ser. No. MT-28666, removal of input shaft from transmission requires removal of the output shaft. On later models, the input shaft can be removed without disturbing the output shaft.

POWER (PTO) SHAFT AND BEARINGS. Bearings for the power shaft (19) can be adjusted after detaching the transmission from the center frame as in paragraph 71. Adjustment is accomplished by adding or removing shims under shaft cover to obtain 0.003 preload. The power shaft can be removed from transmission after removing both the output and input shafts.

R & R AND OVERHAUL

71. **DISCONNECT TRANSMISSION FROM CENTER FRAME.** To detach transmission and final drives assembly from center frame (clutch housing), proceed as follows: Disconnect brake rods. On tractors equipped with Touch-O-Matic, disconnect oil lines from rock shaft unit. Support both halves of tractor separately and remove nuts and cap screws retaining transmission case to center frame. Slide front half of tractor forward (not more than two inches) on studs as shown in Fig. JD346. Bump coupling pin out of propeller (clutch) shaft (21-Fig. JD345) and transmission input shaft, and move center frame away from transmission.

72. **R & R TRANSMISSION.** To remove transmission from tractor and from final drive units, proceed as follows: Disconnect brake rods. On tractors equipped with Touch-O-Matic, remove oil lines from rockshaft. Support both halves of tractor (transmission and center frame) separately and remove nuts and cap screws retaining transmission case to center frame. Slide front half of tractor forward (not more than two inches) on studs. Bump Groov pin out of propeller (clutch) shaft and transmission input shaft and move center frame away from transmission as shown in Fig. JD346. Support transmission housing

and remove drawbar assembly and rear wheels. Remove fender bracket to final drive housing cap screws. Remove fender, bracket and foot rest from transmission as a unit. On model MT, remove each brake unit. Remove two upper cap screws retaining each final drive to transmission and insert two 6-inch studs in their places. Remove remaining cap screws and slide final drive units off tractor.

73. **OVERHAUL.** Data on overhauling the various transmission components are outlined below.

74. OVERHAUL SHIFTER RAILS. To overhaul rails and forks, remove transmission cover. Slip rubber boot (2—Fig. JD351) off shift lever. Place lever in neutral position and remove snap ring (3), upper bearing (beveled washer) (4) and lever. Place sec-

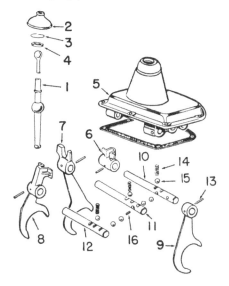

Fig. JD351—M and MT transmission cover and shifter assembly. Ribs inside boot (2) provide air vent for transmission case.

1. Shift lever	10. 3rd shift rail
3. Snap ring	11. 1st & 4th shift rail
4. Lever bearing	12. 2nd & reverse shift rail
5. Cover	
6. 3rd shifter guide	13. Fork pin
7. 1st & 4th fork	14. Detent spring
8. 2nd & reverse fork	15. Detent ball
9. 3rd fork	16. Interlock pin

ond and reverse fork (8) against boss of cover and bump pin out of fork. Use a soft drift and bump rail (12) out of fork and cover. Remove poppet ball (detent ball) and spring and interlocking balls and pin. Follow same procedure with first and fourth speed forks and with third speed shifter and fork. Refer to paragraph 115 for data on repair of belt pulley control.

Length of new interlocking pin (16) is 0.527-0.532. When installing shifter lever boot (2), check to be certain that the boot does not cover the vent which is located in shifter cover.

75. OVERHAUL OUTPUT SHAFT (BEVEL PINION). To remove the output shaft (23—Fig. JD356), which is integral with the main drive bevel pinion, first remove the transmission assembly as outlined in paragraph 72. Also remove the differential as outlined in paragraph 80. Refer also to Fig. JD350.

75A. Remove transmission cover. Remove output shaft bearing cover (31) and nut from forward end of output shaft. Pull shaft rearward and remove front bearing cone and shims (30). Withdraw shaft through rear of case and gears from above. Output shaft is supplied only as a matched pair with the bevel ring gear.

Shims (25) between rear bearing cup (24) and transmission housing wall control mesh position of main drive bevel pinion gear.

When reassembling, place bearing spacer (29) with chamfered edge toward shoulder on shaft. Vary thickness of shims (30) between spacer (29) and front bearing cone to remove all bearing play but permitting shaft to rotate freely; then remove 0.005 thickness of shims to obtain correct bearing pre-load.

Fig. JD352—Removing M and MT second and reverse shifter rail and fork.

6. 3rd shifter guide
7. 1st & 4th fork
8. 2nd & reverse fork
9. 3rd fork
10. 3rd shift rail
11. 1st & 4th shift rail
12. 2nd & reverse shift rail

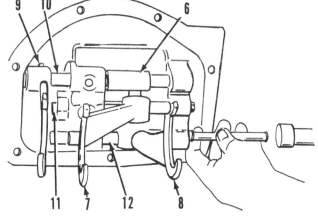

1. Input shaft
4. 1st cluster gear
8. 4th drive gear
10. 3rd drive gear
12. Bearing quill
15. Cluster gear
23. Output shaft
26. 2nd & reverse sliding gear
27. 1st and 4th sliding gear
28. 3rd sliding gear
31. Shaft cover

ing same with cluster gear (15). With draw shaft through front of transsion housing and remove gears, bearings, spacers and coupling from above.

When reassembling install spacer washer (14) between rear bearing cone and cluster gear with beveled edge toward the rear. Vary thickness of shims (21) between bearing cover and transmission wall to remove all shaft play but permitting shaft to rotate freely; then remove shims equal to a 0.003 thickness to obtain correct bearing pre-load of 0.003.

79. OVERHAUL PTO SHIFTER CONTROLS. To remove power shaft shifter control, unscrew plug (12—Fig. JD355) from case and extract poppet (detent) assembly (10 and 11) from left side of case. Remove set screw (8) from shifter fork (6) and remove lever shaft (5) from fork. Top of bushing (4) should extent ¼-inch above top of case. Bushing is presized and, if carefully installed, requires no final sizing.

76. OVERHAUL INPUT SHAFT. To remove the input shaft (1) it is necessary to first remove the transmission and the differential as in paragraphs 72 and 80. On M tractors before Ser. No. M-45073 and MT tractors before Ser. No. MT-28666, the output shaft must be removed before the input shaft can be removed. On later models, the input shaft can be removed without distributing the output shaft. Remove input shaft front bearing quill from transmission front wall. Bump shaft forward through opening in front of housing, and lift rear end of shaft up and out through opening in top of case.

To disassemble input shaft assembly, use a bearing puller and remove front bearing cone. Pull or press third speed drive gear (10) off shaft and remove Woodruff key. Remove spacer (9) and pull fourth speed drive gear (8) off shaft and remove Woodruff

key. Remove spring washer (7), spacer washer (6), cluster gear (4), bearings and rear spacer washer (3).

The spacer washers are installed with chamfered side facing toward rear. Install spring washer (7) with concave (cupped) side nearest plain washer (6). Install fourth speed gear with chamfered end of teeth toward rear and third speed gear with chamfered end of teeth toward front. Vary thickness of shims (11) between front bearing quill (retainer) and case to provide 0.002-0.004 shaft end play. Install oil seal (13) in quill with lip of seal facing toward rear.

78. OVERHAUL POWER (PTO) SHAFT. With output and input shafts removed (as outlined in preceding paragraphs), remove power shaft front bearing cover (22) and shims (21). Bump rear end of shaft forward and remove front bearing cup. Remove rear bearing cone from shaft by bump-

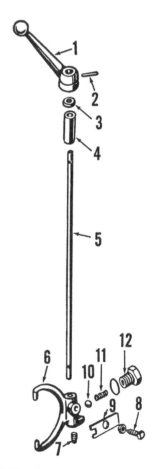

Fig. JD355—M and MT power (PTO) shaft shifter assembly. Presized bronze bushing (4) extends ¼ inch above top of case.

1. Lever	7. Plug
2. Pin	8. Set screw
3. Felt	9. Lock plate
4. Bushing	10. Poppet ball
5. Lever shaft	11. Poppet spring
6. Shifter fork	12. Plug

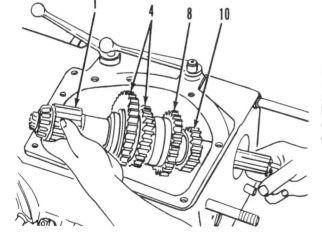

Fig. JD354 — Removing Deere M and MT transmission input shaft (1). First cluster gear (4), 4th drive gear (8) and 3rd drive gear (10) are removed with the shaft.

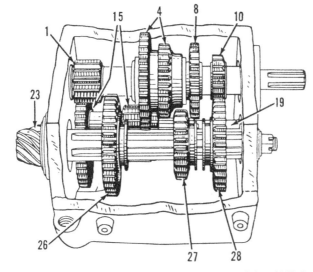

1. Input shaft
4. 1st cluster gear
8. 4th drive gear
10. 3rd drive gear
15. Power shaft cluster gear
19. Power shaft
26. 2nd & reverse sliding gear
27. 1st & 4th sliding gear
28. 3rd sliding gear

Fig. JD356—Cutaway view of M and MT transmission. Main drive bevel pinion (23) is integral with output shaft.

FINAL DRIVE

DIFFERENTIAL

80. **REMOVE & REINSTALL.** To R&R the differential, first remove the final drive assemblies as outlined in paragraph 95 or 100. Remove belt pulley unit or rear cover plate, and on model M, remove both brake assemblies. Remove both differential bearing quills (1—Figs. JD360 and 361). Working through transmission housing rear opening, remove bevel ring gear and differential unit.

On model MT, shims (4) interposed between bearing cones (3) and differential case (9) control differential bearing adjustment and backlash of main drive bevel ring gear and pinion.

On model M, shims (15) interposed between bearing quills (1) and transmission case control bearing adjustment and backlash of main drive bevel ring gear and pinion. Recommended bearing adjustment is 0.002-0.005 pre-load; recommended backlash is 0.006-0.008. Refer to MAIN DRIVE BEVEL GEARS section for adjustment procedure. Install oil seals (2) in quills with lip edge of seal facing differential.

81. **OVERHAUL.** Neither the bevel pinion nor ring gear are furnished separately, but only as a matched pair. Bevel ring gear (5) is attached to differential case by cap screws. To disassemble the differential unit, remove bevel ring gear and bump pinion shaft retaining (Groov) pin (10) out of differential case. Remove differential pinion shaft, pinions, side gears and thrust washers.

Reassemble in reverse order and stake differential case to prevent pinion shaft retaining (Groov) pin from backing out. When new case is installed on model MT, obtain initial bearing adjustment as follows: Insert shims (4) amounting to a thickness of 0.025 between left bearing cone and differential case. For the right bearing cone, insert shims amounting to a thickness of 0.040. Differential carrier bearings are adjusted by varying the amount of shims to provide a 0.002-0.005 pre-load. Shims (4) also control the ring gear to main drive bevel pinion backlash adjustment.

After reassembly, check trueness of ring gear back face by mounting the unit in its normal operating position. Total run-out should not exceed 0.003.

MAIN DRIVE BEVEL GEARS

90. **GEAR MESH.** The main drive bevel pinion output shaft fore and aft position in the transmission case is adjustable (by means of shims) to provide the correct cone center distance or mesh position of the pinion with the main drive bevel ring gear. To adjust the bevel pinion mesh position, it is necessary to first remove the transmission from the tractor and then remove the differential unit.

The main drive bevel gears are available only as a matched pair. If new gears are being installed or if the original gears require adjustment do so as outlined in the following paragraphs.

91. **SETTING THE PINION.** With differential removed and transmission removed from tractor, first step in setting the pinion position is to make certain that pinion (output) shaft bearings are correctly adjusted to 0.005 bearing pre-load. To make this adjustment, remove bearing cover at front of transmission case and nut from front end of pinion shaft. Bump pinion shaft rearward until shims (30—Fig. JD350) are accessible. Vary the number of shims until all end play is removed but shaft rotates freely; then remove 0.005 thickness of shims. This

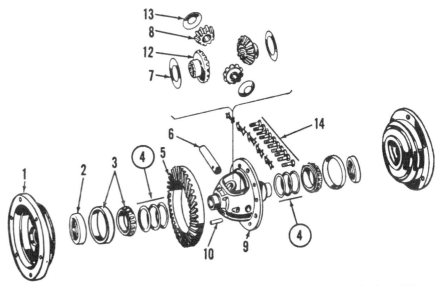

Fig. JD360—Deere MT differential assembly. Bevel ring gear (5) is retained to differential case (9) by cap screws. Shims (4) are for adjustment of differential bearings and backlash of main drive bevel gears.

1. Bearing quill
2. Oil seal
3. Bearing cup & cone
4. Shims (MT)
5. Ring gear
6. Pinion shaft
7. Thrust washer
8. Pinion
9. Differential case
10. Groov pin
12. Side gear
13. Thrust washer

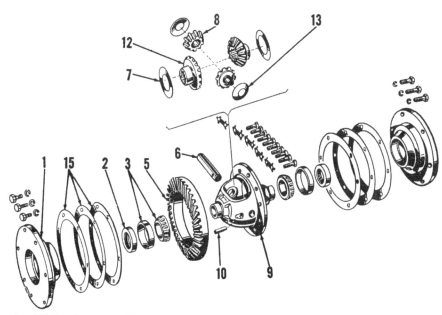

Fig. JD361—Deere M differential assembly. Shims (15) are for adjustment of differential bearings and backlash of main drive bevel gears. Refer to Fig. JD360 for legend.

will provide the desired 0.005 preload of the bearings when the pinion shaft nut is securely tightened.

91A. After bearings end play is adjusted as in paragraph 91, install special pinion setting gage John Deere AM-452-T in transmission case as shown in Fig. JD365.

NOTE: Gage set AM-452-T is for series M tractors and when used for Model MT, it requires two adapter rings which are available from the factory.

Note cone center distance etched on rear face of pinion. Measure the distance between end of gage and bevel pinion rear face with a feeler gage as shown. The thickness of the feeler gage plus 2.782 should equal the cone center distance as etched on the rear face of the pinion. Move the pinion in or out as required by adding or re-

moving shims (25—Fig. JD350), until the feeler gage measurement when added to 2.782 exactly equals the cone distance (such as 2.815) etched on end of pinion. To vary the shims (25) which are located between the rear bearing cup and wall of the transmission housing, it will be necessary to remove the pinion shaft from the transmission.

If John Deere gage is not available, the bevel pinion adjustment can be obtained with an inside micrometer by measuring the distance between the gear end of the bevel pinion and a close fitting mandrel installed in the differential quill bores. In this case, the micrometer reading plus one-half of the mandrel diameter should equal the measurement as indicated on the pinion.

92. BACKLASH ADJUSTMENT. On model MT, vary thickness of shims (4 — Fig. JD360) between bearing cones (3) and differential case (9) to remove all end play and yet permit differential to rotate freely; then remove 0.004 of shims to obtain correct bearing pre-load of 0.002-0.005. After bearings are adjusted, adjust gear backlash to 0.006-0.008 by transferring shims (4) from left side of case to right side of case to increase gear backlash or from right side of case to left side of case to decrease backlash.

Procedure on model M is the same as MT except that backlash and differential bearing shims are located between bearings quills and case as shown in Fig. JD361.

92A. **RENEWAL OF PINION & RING GEAR.** Neither the bevel pinion or the bevel ring gear are sold separately. They must be renewed as a matched pair. To renew the bevel pinion it is first necessary to remove the transmission from tractor and from the final drive units as outlined in paragraph 72, and to remove the differential. The pinion shaft can now be renewed by following the procedure given in paragraph 75A. To adjust mesh position of the new pinion shaft follow the procedure given in paragraph 91A.

The bevel ring gear and differential unit was removed from the tractor in removing the bevel pinion. The bench procedure for removal of the bevel ring gear is self-evident. To readjust the differential carrier bearings and the tooth backlash of the bevel gears, follow the procedure given in paragraph 92.

FINAL DRIVE UNITS

When a final drive unit has been removed for any reason, inspect for oil leakage from the differential. If there are signs of even slight leakage be sure to renew the seal (2) located in differential bearing quill on each side of differential.

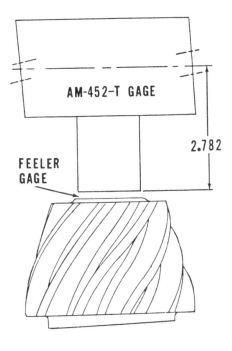

Fig. JD366—The fixed cone center distance 2.782 of the special Deere pinion gage, plus the thickness of the feeler gage required to fill the gap between gage and pinion, should equal the cone center distance as etched on the bevel pinion. Move pinion in or out until this condition is obtained.

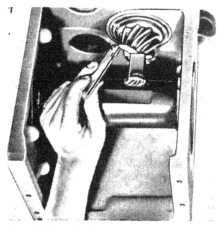

Fig. JD365—Setting the M and MT main drive bevel pinion mesh position with a John Deere special gage AM-452-T. Refer to text for alternate method.

Model MT

93. WHEEL AXLE BEARING ADJUSTMENT. Support rear portion of tractor and remove wheel. Vary the number of shims (18—Fig. JD370) until all end play is removed but shaft is still permitted to rotate freely; then remove approximately 0.004 thickness of shims to obtain a desired pre-load of 0.003-0.006.

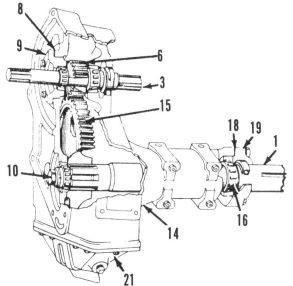

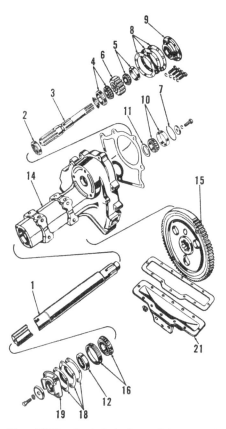

Fig. JD370—Final drive assembly for one side of Deere MT tractor. Shims (8) and (18) are used for adjusting the drive shaft and wheel axle bearings.

1. Wheel axle shaft
3. Drive shaft
6. Drive (bull) pinion
8. Shims
9. Bearing quill
14. Final drive housing
15. Drive (bull) gear
16. Bearing
18. Shims
19. Oil seal quill
21. Housing oil pan

94. DRIVE (PINION) SHAFT BEARING ADJUSTMENT. Remove final drive assembly as outlined in paragraph 95. Vary the number of shims (8) to obtain a pre-load of 0.000-0.003.

95. R&R FINAL DRIVE UNIT. Support rear portion of tractor and remove drawbar assembly and rear wheels. Remove fender bracket to final drive housing cap screws. Remove fender, bracket and foot rest from transmission as a unit. Remove brake unit. Remove the two uppermost cap screws which retain final drive housing to transmission housing and insert two long studs (about 6 inches) in their place. Using a hoist, slide final drive unit off studs.

96. R&R WHEEL AXLE SHAFT, BEARINGS AND/OR BULL GEAR. Remove the final drive assembly as outlined in paragraph 95. Remove axle shaft quill (19) and inner bearing cap screw (10). With the aid of a reaction pusher, push the shaft toward the wheel end and out of bull gear (15). Inner bearing cup is retained in position in the final drive housing by a snap ring.

97. R&R DRIVE SHAFT, BEARINGS AND/OR BULL PINION. Remove final drive assembly from tractor. Shaft, bearings, and bull pinion (6) can be bumped out after removing bearing quill (9).

Bearing cup and/or oil seal can be renewed at this time. Install new seal with the lip facing inward.

When reinstalling the drive shaft assembly, vary the number of shims (8) to obtain the desired bearing pre-load of 0.000-0.003.

Model M

98. WHEEL AXLE ADJUSTMENT. Support rear of tractor. Vary thickness of shims (10—Fig. JD374) between bearing cover (11) and housing to remove all play but permitting shaft to turn freely; then remove approximately 0.004 thickness of shims to obtain the desired pre-load of 0.003-0.006. Tighten inner bearing cap screws to a torque of 180-190 ft.-lbs.

99. DRIVE (PINION) SHAFT BEARING ADJUSTMENT. Support rear of tractor and remove wheel. Remove rear cover or belt pulley unit to check shaft end play. Remove quill

Fig. JD373 — Cut-a-way view of Deere M final drive.

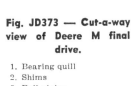

1. Bearing quill
2. Shims
5. Bull pinion
6. Bearing
7. Bull pinion shaft
8. Oil seal
9. Housing
15. Axle
16. Bull gear
18. Spacer
19. Bearing
20. Oil seal
21. Bearing quill

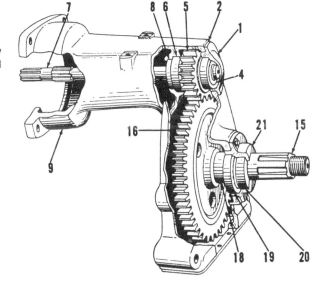

Fig. JD371—Exploded view of Deere MT final drive. Shims (8) and (18) are used for adjusting the drive shaft and wheel axle bearings.

1. Wheel axle shaft	11. Washer
2. Oil seal	12. Oil seal
3. Drive shaft	14. Final drive
4. & 5. Bearing	housing
6. Drive (bull) pinion	15. Drive (bull) gear
7. Snap ring	16. Bearing
8. Shims	18. Shims
9. Bearing quill	19. Oil seal quill
10. Bearing	21. Housing oil pan

(1-Fig. JD374) from outer bearing. Vary thickness of shims (2) to provide from 0.001 pre-load to 0.002 end play in shaft.

100. R & R ONE FINAL DRIVE UNIT. Support rear of tractor and remove wheel. Disconnect brake rod, remove clevis and locknut from rod and push rod out through rear of housing. Detach fender from housing. Remove cap screws attaching final drive to housing and remove assembly. Reinstall in reverse order and adjust brake as in paragraph 112.

101. R & R WHEEL AXLE SHAFT, BEARINGS AND/OR BULL GEAR. Support rear of tractor and remove wheel. Remove the housing oil pan and outer bearing quill (21). Remove inner bearing cover (11) and cap screw and washers from inner end of wheel axle. Drive or press axle out of inner cone and gear and out through outer opening in housing. When shaft is being driven or pressed out, support hub of gear with a length of 3½ inch pipe slipped over outer end of axle. Reassemble in reverse order and adjust bearings as in paragraph 98. Install oil seal (20) in quill with sharp edge of seal towards gear.

102. R & R DRIVE SHAFT, BEARINGS AND/OR BULL PINION AND OIL SEAL. Support rear of tractor and remove wheel. Remove quill (1-Fig. JD374) from outer bearing and withdraw shaft, bearings and gear. Gear and bearing cones can be pressed off shaft after removing snap ring (3). Inner bearing cup can be removed with a puller. Install oil seal in final drive housing with sharp edge of seal out towards wheel. Adjust bearings as outlined in paragraph 99.

BRAKES
Model MT

110. ADJUSTMENT. Support rear portion of tractor, turn adjusting nut (11—Fig. JD381) in until a slight drag is obtained when turning wheel; then retract nut approximately 3 turns. Follow the same procedure for the other brake. Both brakes should be free when pedal is depressed 1½ inches, but should lock the wheel when pedal is depressed approximately 2½ inches. Synchronize brake pedals by varying length of individual pedal rods.

Fig. JD380—Deere MT disc type brake. Driven (lined) discs (3) are splined to outer end of final drive (bull pinion) shaft. See Fig. JD381 for legend.

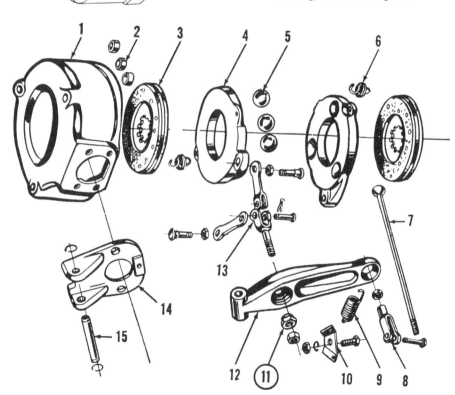

Fig. JD374—Exploded view of Deere M final drive.

1. Bearing quill	12. Washer
2. Shims	13. Bearing
3. Snap ring	14. Washer
4. Bearing	15. Axle
5. Bull pinion	16. Bull gear
6. Bearing	17. Gasket
7. Bull pinion shaft	18. Spacer
8. Oil seal	19. Bearing
10. Shims	20. Oil seal
11. Bearing cover	21. Bearing quill

Fig. JD381—Deere MT disc type brake.

1. Brake housing	6. Release spring
2. Hollow dowels	7. Pedal rod
3. Driven (lined) discs	8. Pedal rod yoke
4. Actuating discs	9. Lever return spring
5. Actuating balls	10. Return spring clip

11. Adjusting nut
12. Operating lever
13. Actuating rod & yokes
14. Operating lever bracket
15. Operating lever pin

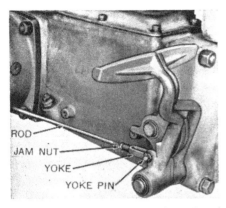

Fig. JD383—Deere M right brake rod and adjusting yoke.

111. **R & R AND OVERHAUL.** Remove the three long cap screws which retain fender bracket and brake housing (1). Disconnect brake pedal rod at pedal shaft. Remove cap screws retaining fender and foot rest to tractor and remove these as a unit. Remove housing and brake assembly.

The brake assembly will be disassembled after removing adjusting nut (11). The two brake actuating

discs (4) are interchangeable as are the two lined discs (3). Renew brake release springs (6) if they have lost their tension. Renew actuating discs (4) and/or brake housing (1) if friction surface is excessively worn. Reinstall assembly and adjust as per preceding paragraph.

Model M

112. **ADJUSTMENT.** Loosen jam nut on brake rods and screw rod into rod yoke until all play is removed from pedal, then shorten rod a half turn. Refer to Fig. JD383.

113. **R & R AND OVERHAUL.** Remove final drive assembly as in paragraph 100. Remove cap screws retaining brake assembly to transmission case and remove assembly and friction disc (8-Fig. JD384). To disassemble, compress springs (2) and remove retaining washers (1). Clean thrust bearing (6) in oil and drain; do not lubricate. Spring pressure when compressed to ¾ inch length should be 80-90 pounds. Reassemble and rein-

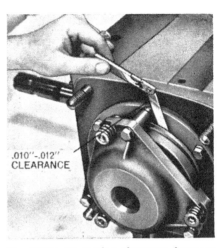

Fig. JD385—Measuring clearance between Deere M lined brake disc and the differential bearing quill.

stall in reverse order. Vary thickness of shims (4), on cap screws, between power plate and differential quill to obtain 0.010-0.012 clearance between friction disc and differential quill. Vary shims on each cap screw to obtain the same clearance all around the disc.

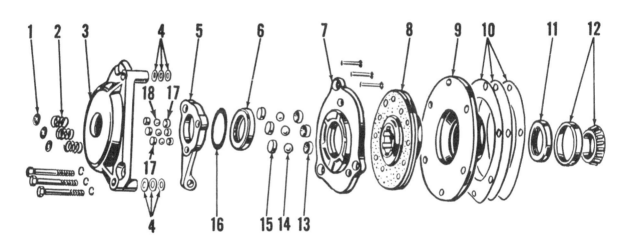

Fig. JD384—Exploded view of Deere M brake. Shims (10) are available in thicknesses of 0.002, 0.005 and 0.010.

1. Washer	5. Actuating lever	8. Lined disc	11. Oil seal	14. Ball
2. Release spring	6. Thrust bearing	9. Differential bearing quill	12. Bearing	16. Spring washer
3. Power plate	7. Primary disc	10. Shims	13. & 15. Ball seat	17. Ball seat
4. Shims (0.002, 0.005 & 0.010)				18. Ball

BELT PULLEY UNIT

The bearing end play of the pulley shaft can be adjusted and the pulley shaft oil seal can be renewed without removing the belt pulley unit from the tractor. Other repairs necessitate the removal of the belt pulley unit.

114. **ADJUST PULLEY SHAFT BEARINGS.** To adjust the end play of the pulley shaft in its bearings, re-

move the nut from end of pulley shaft, and using a suitable puller, remove pulley. Remove the bearing cone (20-Fig. JD390) and vary the shims (19) until shaft rotates freely without perceptible end play. With this condition obtained, remove 0.001-0.002 of shims to obtain the desired 0.000-0.003 preload.

114A. **R&R AND OVERHAUL.** Remove master shield and cap screws retaining unit to transmission.

Remove nut from end of pulley shaft, insert puller in tapped holes of pulley and remove pulley. Remove drive shaft bearing quill (retainer) (2-Fig. JD390) and shims (3). Withdraw drive shaft, gear and inner bear-

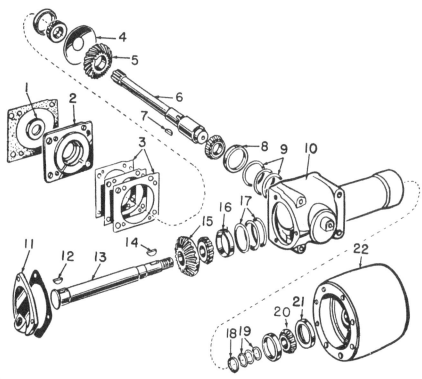

Check gear backlash of assembled unit. If backlash is less than 0.005 or more than 0.010 it should be brought within these limits by disassembling the unit and varying the shims (17) located between bearing cup (16) and housing. If backlash is measured on splines at front end of drive shaft as shown in Fig. JD392 instead of at the gear teeth, the dial indicator reading should be within the limits of 0.0015-0.003 due to the smaller radius at point of measurement.

If the gear heel faces are within 0.005 of being flush with each other after correct backlash has been obtained, the mesh position is satisfac-

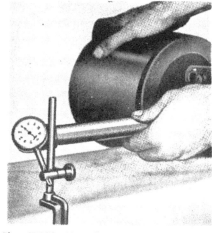

Fig. JD390—M and MT belt pulley unit. Shims (3) and (19) are for bearing adjustment. Shims (9) and (17) are for mesh and backlash adjustment of bevel gears.

1. Oil seal	8. Bearing cup	15. Driven gear
2. Drive shaft quill	9. Gear shims	16. Inner bearing cup
3. Bearing shims	10. Case	17. Gear shims
4. Oil slinger	11. Gear case cover	18. Washer
5. Drive gear	12. Woodruff key	19. Bearing shims
6. Drive shaft	13. Pulley shaft	20. Outer bearing
7. Woodruff key	14. Woodruff key	21. Pulley shaft seal

Fig. JD392—M and MT belt pulley gear backlash cannot be measured at the gear teeth. To measure gear backlash, use a dial indicator with the button of the indicator contacting the splines of the drive shaft. When measured in this manner, backlash reading should be 0.0015-0.003. Gear backlash is controlled with shims (17 —Fig. JD390) located behind the pulley shaft inner bearing cup.

ing cones. Remove cover (11) and bump pulley shaft, gear and inner bearing cone out through left side of case.

If cups for the roller bearing cones are defective they can be removed with pullers otherwise they should not be disturbed. If cups are removed, be sure to avoid damaging the shims (9) and (17) which control the mesh and backlash position of the bevel gears. In most every case, the mesh position will be correctly maintained if the original shims or new ones of same thickness are installed.

114B. During reassembly, observe the following points: Assemble the pulley shaft to the housing before assembling the drive shaft to the housing. Shims (17) control the backlash of bevel gears. If these shims were removed during disassembly, be sure to reinstall the same ones or new ones of exactly the same thickness. Vary the shims (19) until shaft rotates freely without perceptible end play then remove 0.001-0.003 of shims to obtain desired pre-load of 0.002.

If shims (9) were removed during disassembly, be sure to reinstall the same ones or new ones of the same thickness. These shims control the

mesh position of the bevel gear on the driveshaft (6). Vary the shims (3) located between quill and housing until shaft rotates freely without perceptible end play, then add an additional 0.003 of shims to obtain the desired end play of 0.002-0.004.

Fig. JD391 — Checking mesh position of belt pulley drive gear with feelers and John Deere gage number AM-454-T.

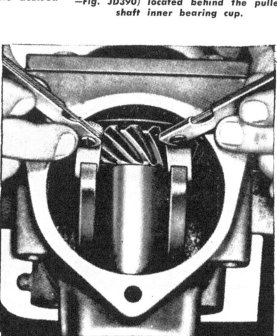

tory. If heel faces are not within 0.005 of being flush, it will be necessary to change the position of the drive gear by disassembling the unit and varying the shims (9). The John Deere Company provides a special gauge AM-454-T which saves considerable time in the job of setting the mesh position and backlash of the belt pulley bevel gears.

115. **R & R CONTROL.** Remove transmission cover and extract detent ball plug (10—Fig. JD393) and spring (9) from hole over pulley shifter yoke. Bump pin (2) out of control lever and remove pin connecting shifter (4) to shifter yoke and release shifter. Remove differential unit as outlined in paragraph 80 and withdraw shifter yoke (7) and coupling (5) through rear opening.

POWER TAKE-OFF

116. The power take-off shaft is part of the transmission assembly and extends through the rear of transmission case. Refer to paragraph 78, in TRANSMISSION SECTION for removal and reinstallation.

POWER LIFT UNIT

The maintenance of absolute cleanliness of all parts is of utmost importance in the operation and servicing of the hydraulic system. Of equal importance is the avoidance of nicks or burrs on any of the working parts.

PUMP

Models M-MT

The two section pump (Fig. JD401) as used on model M tractors before Ser. No. 32884 is not interchangeable with the three section type (Fig. JD-400) as used on later model M tractors and all model MT tractors. The service procedures for the two pumps are similar, and any differences will be indicated in the following paragraphs.

117. **REMOVE & REINSTALL.** The gear type pump is mounted on the engine timing gear cover and receives its drive through coupling (1-Figs. JD400 & 401) which is bolted to the camshaft. To remove pump unit, remove hood and grille, and disconnect oil lines at pump. Remove the three socket head screws retaining pump to timing gear cover and remove pump.

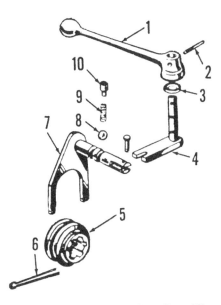

Fig. JD393—M and MT belt pulley shifter control is located in transmission case.

1. Shifter lever	6. Cotter pin
2. Pin	7. Shifter yoke
3. Felt	8. Poppet ball
4. Shifter	9. Poppet spring
5. Coupling	10. Detent plug

118. **OVERHAUL.** Disassemble pump unit by removing the socket head cap screws. Drive shaft (2) can be pressed out of pump gear (9). Needle bearings (6 & 16) can be pressed out for renewal. However, needle bearings (13 & 18) in pump cover must be collapsed to remove same. Pump drive shaft oil seal (3) is of the two way design, and can be installed with either seal lip facing the pump gears.

On two-section pumps, make certain that mating surfaces of pump body and cover are clean and unmarred. Do not use sealing compound or gaskets on these mating surfaces. On three-section pumps make certain that sealing rings (8) are seated in grooves of center section.

If drive shaft and gears do not rotate freely after pump is assembled, then loosen the assembly screws and retighten alternate screws evenly while rotating pump drive shaft until all screws are tightened to a torque value of 24-28 ft. pounds.

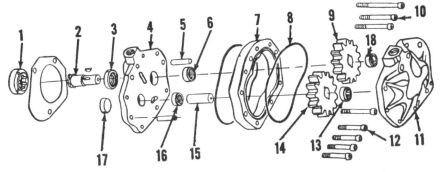

Fig. JD400—Deere M (after Ser. No. 32883) and MT hydraulic pump assembly which is mounted on the engine timing gear cover and receives its drive from the camshaft.

1. Drive coupling	6. Needle bearing	14. Driven gear
2. Drive shaft	7. Center section	15. Driven gear shaft
3. Double seal	8. Pump seals	16. Needle bearing
4. Pump base	9. Drive gear	17. Expansion plug
5. Dowel pins	13. Needle bearing	18. Needle bearing

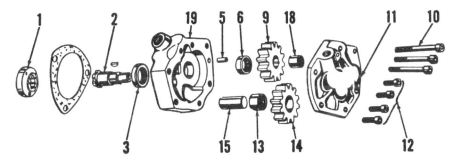

Fig. JD401—Deere M (before Ser. No. 32884) hydraulic pump. This two section pump receives its drive and operates in a similar manner to that shown in Fig. JD400.

1. Drive coupling	9. Drive gear	13. Needle bearing
2. Drive shaft	10. Mounting screws	14. Driven gear
3. Double seal	11. Pump cover	15. Driven gear shaft
5. Dowel pin		18. Needle bearing
6. Needle bearing		19. Pump body

On the MT, repeat the preceding procedure when checking and adjusting the other relief valve. Be sure to remove the ⅜-inch pipe plug after the adjustment of each valve is completed and reinstall the ½-inch pipe plug.

120. QUADRANT—ADJUST. To adjust the quadrant, start engine and pull control lever rearward. With the rock shaft lift arms at the top of their travel, the right-hand lift arm should have a clearance of ⅛ to $\frac{3}{16}$-inch between lift arm and seat frame shield. To obtain this clearance, rotate the quadrant on its two attaching studs.

121. R&R VALVE HOUSING. To perform any overhaul work on the valve housing it is advisable to remove the valve housing and rock shaft assembly as a unit. Disconnect the pressure lines and suction lines at instrument panel, valve housing and rockshaft housing. NOTE: The suction line extends into rockshaft housing approximately 4 inches. Do not bend or damage this line during removal. Remove tractor seat and on model M, remove quadrant. Remove cap screws retaining valve housing to transmission housing and lift Touch-O-Matic unit from tractor.

After unit is removed from tractor, remove cap screws retaining rock shaft housing to valve housing, and separate the two units far enough to remove "hairpin" retainers from Groov pins (11—Fig. JD420 or 421) on crank arms. Slip equalizer rods from Groov pins and remove rock shaft housing. Remove cap screws retaining front cover to valve housing and separate the two units.

122. PISTON SEALS. With the valve housing removed as outlined in paragraph 121, remove ½-inch plugs which are installed at point (A—Fig. JD405 or 406), and use low pressure air to carefully force piston (18) out of cylinders.

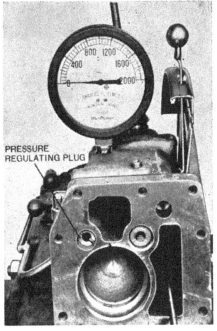

Fig. JD409—Checking the relief valve operation on Deere M valve housing. Refer to caption for Fig. JD408.

PRESSURE REGULATING PLUG

Fig. JD411—Deere MT. Lift unit valve housing showing locations of relief valves. To adjust relief valves, separate rock shaft housing from valve housing, and remove adjusting screw lock screw; then, turn slotted adjusting screw in to increase pressure or out to decrease pressure.

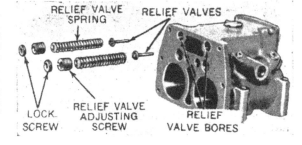

RELIEF VALVE SPRING — RELIEF VALVES

LOCK SCREW — RELIEF VALVE ADJUSTING SCREW — RELIEF VALVE BORES

Fig. JD412 — Relief valve exploded from Deere M valve housing. Refer to caption for Fig. JD411. Late models are equipped with ball type relief valves.

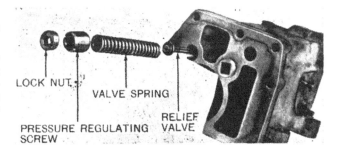

LOCK NUT — VALVE SPRING

PRESSURE REGULATING SCREW — RELIEF VALVE

Mark all mating parts for proper reassembly. Install seals (17) (leather ring and neoprene ring) so that lip of seal faces piston head (6) (front of tractor).

123. CYLINDER HEADS & SEALS. Remove pistons (18) as outlined in preceding paragraph. Pry out snap rings (7). Place valve housing in a press and push cylinder heads (6) out of housing. Always install new head seals when reassembling.

Do not remove cylinder head unless a new valve housing, cylinder head and/or head seal is to be installed.

124. CHECK VALVES. Check valves (3) may be removed after removing the retaining plugs and springs. Also remove shims (4) and push rods (5). Mark all mating parts for correct reassembly.

Free length of check valve springs should be $2\frac{1}{16}$ inches. Renew springs if they do not test 3.8-4.6 pounds when compressed to a height of ¾-inch. Seat

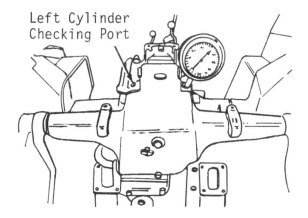

Left Cylinder Checking Port

Fig. JD408 — Deere MT. Checking operation of one relief valve (right-side) by removing a ½ inch pipe plug and installing a ⅜ inch pipe plug directly into the threaded passage beneath the ½ inch plug just removed; then, install a pressure gage (1000 psi capacity) where the ½ inch plug was removed. Refer to paragraph 119. Check operation of relief valve on left-side in a similar manner.

angle on the check valves is 45 degrees, and a damaged valve can be refaced on a machine and reseated by hand lapping.

To adjust the check valve or valves (3), it is necessary to first install the control valve or valves (15) with tapered ramp facing towards the check valve bore in the housing. Correct check valve adjustment is when the by-pass port is uncovered approximately 0.016-0.025 inch by the control valve when the check valve is seated, and the check valve push rod

Check the ground surface of shuttle valve for burrs, sharp edges, scratches and/or any roughness which will prevent valve from operating freely. These conditions may produce a difference in the start and/or speed of lift of one side as compared to the other. Roughness and/or burrs can be removed with Crocus cloth or a fine hone.

ROCK SHAFT HOUSING

130. **R&R AND OVERHAUL.** Rock shaft housing and assembly can be removed without disturbing the valve housing. Remove tractor seat and support rock shaft housing. Remove bolts retaining rock shaft housing to valve housing. Separate the two housings far enough to remove "hairpin" retainers from Groov pins (11—Fig. JD-420 or 421). Disconnect equalizer rods and remove rock shaft housing.

To disassemble the MT unit, proceed as follows: Remove lift arms (5 and 9). To loosen rock shafts in housing, bump on right end of rock shaft, viewed from rear of unit, until shafts are two inches out of housing. Remove hollow shaft (3). Pull inner shaft (1) from left side of housing until spline is out of right crank arm (10). Continue to pull shaft until splines contact left crank arm. Align blind spline of inner shaft with that of the left crank arm and separate the shaft from the arm by pulling shaft through crank arm and left side spacing washer (12).

The bushings (2) can be removed by splitting same with a chisel. Service bushings are presized and should be installed with a suitable mandrel.

To disassemble the M unit, proceed as follows: Remove lift arms (5 and 9 —Fig. JD421). Drive out the two Groov pins which retain the crank arm to the rockshafts, and using a suitable puller, pull rockshafts from crank arm and housing. Drive out the Groov pin which retains the piston rod to the crank arm. Rockshaft seals can be renewed at this time.

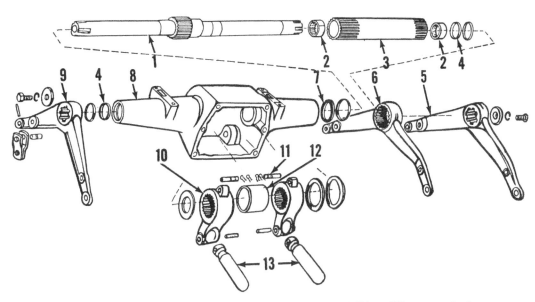

Fig. JD420—Deere MT lift unit rockshaft, exploded view. Bushings (2) are pre-sized.

1. Rockshaft	5. Left lift arm	8. Rockshaft housing	11. Crank arm pins
3. Hollow rockshaft	6. Inside arm	9. Right lift arm	12. Crank arm spacer
4. Seal and retainer	7. Rockshaft seal	10. Crank arms	13. Connecting rods

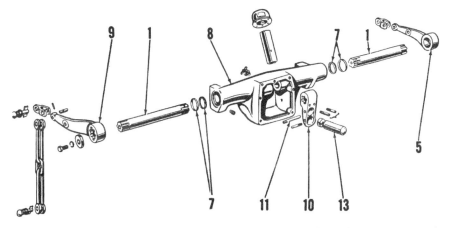

Fig. JD421—Exploded view of Deere M rockshaft housing and associated parts as used on Ser. No. M10001-M46204. Late models are equipped with a one piece rockshaft.

1. Rockshaft	8. Rockshaft housing	11. Crank arm pin
5. Left lift arm	9. Right lift arm	13. Connecting rod
7. Seal and retainer	10. Crank arm	

Field draw
Volts 6.0
Amperes 2.5-2.72
Output (Cold)
Volts 7.0
Amperes 20-25
Rpm 2400

Generator Model 1100955
Brush spring tension..........16 oz.
Field draw
Volts 12.0
Amperes 2.0-2.14
Output (Cold)
Volts 14.0
Amperes 11-13
Rpm 2300

**Generator Models, 1101356, 1101371,
1101377 and 1101385**
Brush spring tension..........16 oz.
Field draw
Volts 6.0
Amperes 3.5-4.5
Output (Cold)
Volts 7.5-7.8
Amperes 11-13
Rpm 1800

Generator Model 1101390
Brush spring tension..........24 oz.
Field draw
Volts 6.0
Amperes 2.6-2.9
Output (Cold)
Volts 6.9-7.1
Amperes 15-17
Rpm 2000

Generator Model 1101755
Brush spring tension..........16 oz.
Field draw
Volts 12.0
Amperes 1.6-1.69
Output (Cold)
Volts 14.4-14.9
Amperes 8-10
Rpm 2200

Generator Model 1101777
Brush spring tension..........16 oz.
Field draw
Volts 12.0
Amperes 1.6-1.69
Output (Cold)
Volts 13.8-14.2
Amperes 8-10
Rpm 2000

Generator Model 1101857
Brush spring tension..........16 oz.
Field draw
Volts 6.0
Amperes 2.61-3.0

Output (Cold)
Volts 7.0
Amperes 10-13
Rpm 2400

**Regulator Models 1116809
and 1116816**
Cut-out relay
Air gap, inches............ 0.020
Point gap, inches.......... 0.022
Closing voltage (range).....6.4-7.8

Regulator Model 1116810
Cut-out relay
Air gap, inches............ 0.015
Point gap, inches.......... 0.020
Closing voltage (range)...13.0-14.5

**Regulator Models 1118265
and 1118305**
Cut-out relay
Air gap, inches............ 0.020
Point gap, inches.......... 0.020
Closing voltage (range).....5.9-7.0
Adjust to 6.4
Voltage regulator
Air gap, inches............ 0.075
Voltage range 6.6-7.2
Adjust to 6.9

**Regulator Models 1118266
and 1118306**
Cut-out relay
Air gap, inches............ 0.020
Point gap, inches.......... 0.020
Closing voltage (range)...11.8-14.0
Adjust to 12.8
Voltage regulator
Air gap, inches............ 0.075
Voltage range 13.6-14.5
Adjust to 14.0

Regulator Model 1118786
Cut-out relay
Air gap, inches............ 0.020
Point gap, inches.......... 0.020
Closing voltage (range).....5.9-6.7
Voltage regulator
Air gap, inches............ 0.075
Voltage range 6.8-7.4

Regulator Model 1118792
Cut-out relay
Air gap, inches............ 0.020
Point gap, inches.......... 0.020
Closing voltage (range)...11.8-14.0
Voltage regulator
Air gap, inches............ 0.075
Voltage range 13.6-14.5

**Starting Motor Models 760
and 1108914**
Brush spring tension (oz.).......36-40
No load test
Volts 5.65

Maximum amperes 70
Rpm 3000
Lock test
Volts 3.0
Maximum amperes 500
Torque, ft.-lbs. 19

**Starting Motor Models 1107424
and 1107445**
Brush spring tension (oz.).....24 min.
No load test
Volts 5.0
Maximum amperes 65
Rpm 5000
Lock test
Volts 3.15
Maximum amperes 65
Torque, ft.-lbs. 15

Starting Motor Model 1107942
Brush spring tension (oz.).....24 min.
No load test
Volts 5.7
Maximum amperes 60
Rpm 6000
Lock test
Volts 3.0
Maximum amperes 600
Torque, ft.-lbs. 16

**Starting Motor Models 1108908
and 1108919**
Brush spring tension (oz.).......36-40
No load test
Volts 11.2
Maximum amperes 80
Rpm 4500
Lock test
Volts 5.35
Maximum amperes 670
Torque, ft.-lbs. 32

Starting Motor Model 1108950
Brush spring tension (oz.).......36-40
No load test
Volts 11.3
Maximum amperes 65
Rpm 5500
Lock test
Volts 4.0
Maximum amperes 675
Torque, ft.-lbs. 30

Starting Motor Model 1109600
Brush spring tension (oz.).......24-28
No load test
Volts 5.7
Maximum amperes 60
Rpm 5000
Lock test
Volts 3.75
Maximum amperes 450
Torque, ft.-lbs. 8.5

NOTES

NOTES

NOTES

NOTES

NOTES